FM 55-150

WAR DEPARTMENT FIELD MANUAL

AMPHIBIAN
TRUCK COMPANY
FIELD MANUAL

BY **WAR DEPARTMENT · 15 SEPTEMBER 1944**

DISCLAIMER:

This manual is sold for historic research purposes only, as an entertainment. It contains obsolete information and is not intended to be used as part of an actual operation or maintenance training program. No book can substitute for proper training by an authorized instructor.

WAR DEPARTMENT FIELD MANUAL
FM 55-150

This manual supersedes FM 55-150, 11 October 1943, including C1,
15 November 1943.

AMPHIBIAN
TRUCK COMPANY

WAR DEPARTMENT · 15 SEPTEMBER 1944

United States Government Printing Office
Washington 1944

WAR DEPARTMENT,
WASHINGTON 25, D. C., 15 September 1944.

FM 55–150, Amphibian Truck Company, is published for the information and guidance of all concerned.

[A. G. 300.7 (20 Jul 44).]

BY ORDER OF THE SECRETARY OF WAR:

G. C. MARSHALL,
Chief of Staff.

OFFICIAL:

J. A. ULIO,
Major General,
The Adjutant General.

DISTRIBUTION:

As prescribed in paragraph 9a, FM 21–6; PE (Tng Div) (5); IC 55 (25), [1](50), [2](75), [3](150).

IC 55: T/O & E 55–500, Amphibian Truck Sec Type No. 1 (25); [1]55–500, Amphibian Truck Sec Type No. 2 (50); [2]55–500, Amphibian Truck Sec Type No. 3 (75); [3]55–37 (150).

For explanation of symbols, see FM 21–6.

CONTENTS

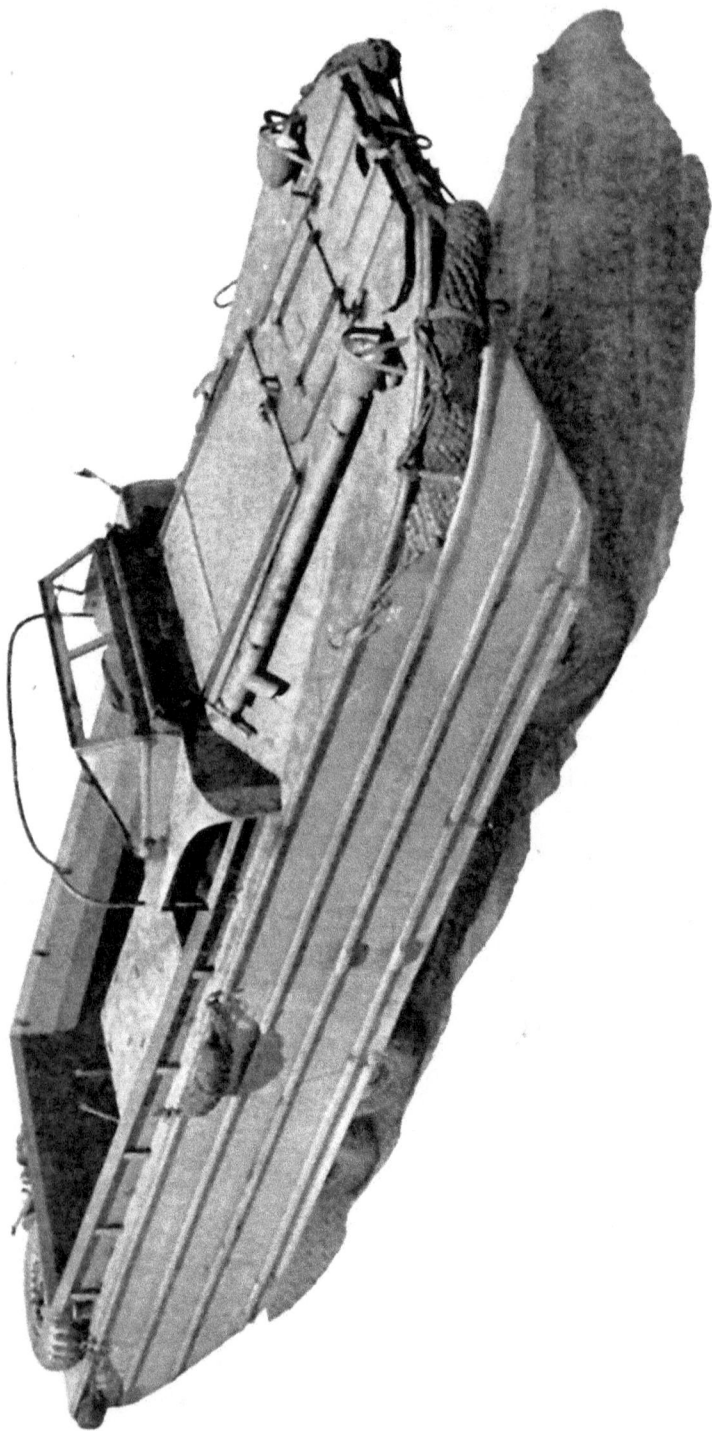

Figure 1. 2½-ton Amphibian Truck, 6 x 6, Model DUKW 353.

This manual supersedes FM 55–150, 11 October 1943, including C1, 15 November 1943.

INTRODUCTION

---

1. PURPOSE AND SCOPE OF MANUAL. This manual explains the organization and operation of the Transportation Corps amphibian truck company. Its purpose is to assist officers and enlisted men of the company in training and operation, and to guide command and staff officers charged with their employment.

No attempt is made to prescribe use of amphibian truck companies in all situations. Since actual conditions in which an amphibian truck company may be utilized vary widely, it is not intended that a literal interpretation of these provisions be made. Ultimate responsibility for operation rests upon the commanding officer, and he must work out details and use of personnel and equipment as the situation demands.

For detailed information on mechanical operation, lubrication, and maintenance of the DUKW, see TM 9–802, and War Department Lubrication Order #505.

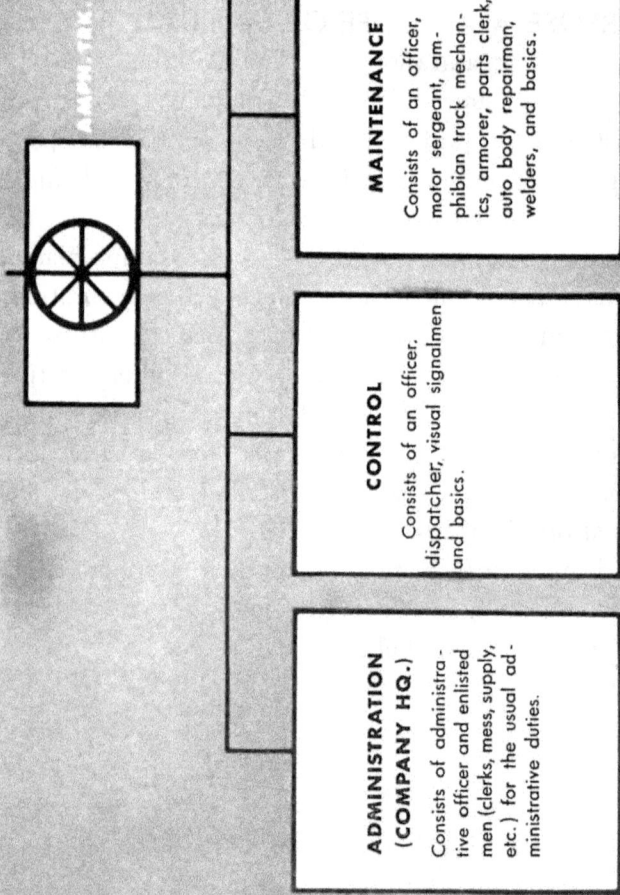

FUNCTIONAL CHART AMPHIBIAN TRUCK COMPANY

AMPH. TRK.

ADMINISTRATION (COMPANY HQ.)
Consists of administrative officer and enlisted men (clerks, mess, supply, etc.) for the usual administrative duties.

CONTROL
Consists of an officer, dispatcher, visual signalmen and basics.

MAINTENANCE
Consists of an officer, motor sergeant, amphibian truck mechanics, armorer, parts clerk, auto body repairman, welders, and basics.

OPERATIONS
Consists of an officer, amphibian truck drivers, crane operator and basics.

Figure 2.

SECTION I

MISSION, ORGANIZATION, AND DUTIES

---

2. MISSION. The mission of an amphibian truck company is to aid the supply and evacuation operations of any force where harbor and normal port facilities either are inadequate or totally lacking. In performing this mission, the company will provide power lighterage service in the transfer of troops, impedimenta, and supplies from ships lying offshore directly to beachhead and inland dumps; supply and provision nearby outposts located on other islands or on nearby beaches inaccessible by land to the principal supply points; and evacuate casualties and transfer material from beachhead and inland dumps directly to ships.

3. ORGANIZATION. The amphibian truck company or section is organized as provided in T/O & E 55–37 or T/O & E 55–500, respectively. For functional purposes, the company is organized into four groups: administration, control, maintenance, and operations. (See fig. 2.)

4. DUTIES OF PERSONNEL. The duties of the key personnel of the amphibian truck company are described briefly below:

  a. **Company commander.** The company commander is both the administrator of the company and the director of its operations. He is, therefore, responsible for training the company, for its efficient administration and operation, and for maintaining discipline.

b. Administrative officer. The administrative officer will have charge of the housekeeping (mess, supply, housing) of the company and other administrative duties as the company commander may direct.

c. Control officer. The control officer controls and coordinates the operations of the company as outlined in Section XIV.

d. Maintenance officer. The maintenance officer is in charge of all maintenance personnel, tools, and equipment. He advises the company commander on the performance of first echelon maintenance, and is responsible for second and third echelon maintenance, for the inspection of vehicles and tools, and for supply of maintenance materials.

e. Operations officer. The operations officer is in command of an operating platoon. He is responsible for the training of the platoon, including tactical and technical subjects, and for the performance of first echelon maintenance. He may assist the control officer at beach, shipside, dump, or dispersal area if required.

f. First sergeant. The first sergeant will prepare or supervise the writing of all reports, pay rolls, organizational records and orders, transmit the company commander's orders to the enlisted men, and assist the company commander in maintaining discipline and morale. He will delegate as many duties as possible in connection with records to the company clerk in order to devote most of his time to operations.

g. Motor sergeant. The motor sergeant is the chief mechanic and principal assistant to the maintenance officer. He is responsible for maintenance records, replacement of parts, salvage of parts, and for the care, storage, and

4

dispensing of lubricants, and for the inspection of tools and equipment.

h. Dispatcher. The dispatcher schedules trips for all DUKW's available for duty. He dispatches vehicles to loading positions at ships or dumps and directs them to unloading stations depending upon their load. He keeps records of DUKW movement, including time of arrival and departure from DUKW control center. He carries out any orders of the DUKW control officer.

i. Platoon sergeant. The platoon sergeant assists in training the platoon and directs its activities in actual operations. He supervises and coordinates the operations of the section and the squad leaders under him, and assumes the duties of the platoon leader in an emergency. He performs independent missions when called upon to do so.

j. Mess sergeant. The mess sergeant supervises the preparation and serving of food, checks the daily menu against items and quantity of subsistence stores, inspects the kitchen, mess hall, and store rooms.

k. Supply sergeant. The supply sergeant issues supplies to personnel, prepares requisitions for supplies and equipment, maintains organizational and individual property records, and maintains stock records of supplies.

l. Section leader. The section leader assists in the training of the section and directs its activities in actual operations. He supervises and coordinates the operations of the squad leaders under him, and assumes the duties of the platoon sergeant in an emergency. He performs independent missions when called upon to do so.

m. Squad leader. The squad leader maintains discipline of the enlisted men of his unit and directs their activities in actual operations. He assumes the duties of the section leader in an emergency.

n. Company clerk. The company clerk assists the first

sergeant and assumes maximum responsibility for handling records and correspondence.

o. Amphibian truck driver. The amphibian truck driver operates the DUKW on land and in the water. He prepares the DUKW for debarkation, supervises loading and helps to unload cargo by hand and with the aid of special equipment. He performs first echelon maintenance as prescribed by TM 9–802, and maintains log of operations. He uses navigational and piloting aids described in Section XI.

p. Amphibian truck mechanic. The amphibian truck mechanic maintains and repairs DUKW's by performing second and third echelon maintenance as prescribed in TM 9–802 and TM 9–1802, A, B, and C.

q. Armorer. The armorer maintains, services, and makes minor repairs on all small arms and crew-served weapons, estimates conditions of weapons and parts and determines whether replacements are necessary. He performs other duties as directed by the maintenance officer.

r. Parts clerk. The parts clerk receives, stores, and issues DUKW equipment and accessories, and spare parts and tools used in connection with operations in the motor pools and repair shops. He maintains stock record cards, and receives, stores, and cares for supplies, such as oil, grease, paint, and cleaning materials.

s. Cook. The cook prepares food for the personnel of of the company, using a daily menu as a guide. He follows army methods of food preparation and all incidental details. When necessary, he sets up and operates a field range.

t. Crane operator. The crane operator operates a power crane which lifts and moves heavy cargo in loading or unloading DUKW's or moves loads from one place to another within the dump.

u. Auto body repairman. The auto body repairman

fabricates, assembles, installs, and repairs the DUKW hull, and makes any other necessary body repairs.

v. Visual signalman. The visual signalman sends and receives messages by visual signaling devices. He transmits messages by International Morse Code or other code using light, sound, wigwag or semaphore.

w. Welder. The welder fuses metal parts and repairs breaks or cracks in hulls and frames of DUKW's or other company equipment. He may also build or improvise special equipment for the operations units to use in the performance of their mission.

SECTION II

WHAT IS THE DUKW?

---

5. DUKW. This manual describes the use of the vehicle known as Truck, Amphibian, 2½-ton, 6 x 6 DUKW–353. The vehicle will be referred to in the manual as the DUKW, which is pronounced "duck." The DUKW is built for operation on land and water and is the standard 2½-ton, 6 x 6 truck made amphibious by the addition of a watertight hull and a water propeller driven by the engine. (See fig. 3.)

6. ITS USES. Situations in which the DUKW may operate to greatest effect are those where port facilities are lacking, where outlying reefs, sandbars, and heavy surf make the use of conventional lighters, landing barges,

Figure 3. DUKW compared to a truck, 2½-ton, 6 x 6.

and other craft impractical, and where the transfer of cargo to beachhead and inland dumps ordinarily would necessitate the use of both watercraft and land vehicles.

7. LAND AND WATER. The character of the DUKW is such that the transfer of men and supplies, involving travel on both land and water, may be made in one continuous operation. The fact that the vehicle can operate equally well on land or water should not be abused. For example, long land hauls should be made by conventional land trucks, and long water hauls by boats, barges, etc.

8. OPERATION. For operating on land, the vehicle has six wheels driven by a conventional six-cylinder engine. The front wheels are steered by a conventional steering gear assembly. In water operations the vehicle is driven by a water propeller and steered by a rudder, both of which are mounted in a tunnel at the rear.

9. DESCRIPTION. a. The vehicle is identified by its six wheels, its boat-shaped body or hull, and its low silhouette cab and cargo compartment.

b. The hull has a deck forward of the driver's compartment, to the rear of the rear wheels, and along both sides. The two-man cab, or driver's compartment, is an open type with a removable canvas top and open back. The driver's and assistant driver's seats are entered from the rear or overside. Cargo space, provided at the rear of the cab, will accommodate approximately 25 men and their equipment or, under normal conditions, approximately 5,000 pounds payload. See figures 4, 5, and 6 for nomenclature.

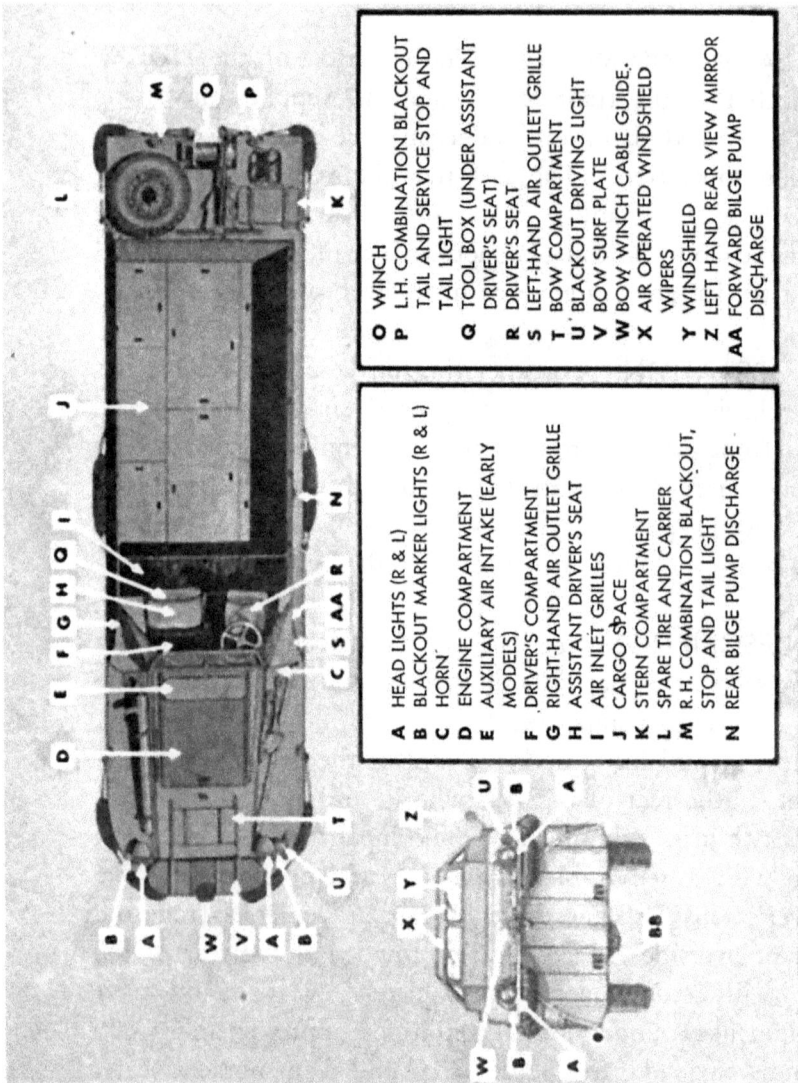

A HEAD LIGHTS (R & L)
B BLACKOUT MARKER LIGHTS (R & L)
C HORN
D ENGINE COMPARTMENT
E AUXILIARY AIR INTAKE (EARLY MODELS)
F DRIVER'S COMPARTMENT
G RIGHT-HAND AIR OUTLET GRILLE
H ASSISTANT DRIVER'S SEAT
I AIR INLET GRILLES
J CARGO SPACE
K STERN COMPARTMENT
L SPARE TIRE AND CARRIER
M R.H. COMBINATION BLACKOUT, STOP AND TAIL LIGHT
N REAR BILGE PUMP DISCHARGE

O WINCH
P L.H. COMBINATION BLACKOUT TAIL AND SERVICE STOP AND TAIL LIGHT
Q TOOL BOX (UNDER ASSISTANT DRIVER'S SEAT)
R DRIVER'S SEAT
S LEFT-HAND AIR OUTLET GRILLE
T BOW COMPARTMENT
U BLACKOUT DRIVING LIGHT
V BOW SURF PLATE
W BOW WINCH CABLE GUIDE.
X AIR OPERATED WINDSHIELD WIPERS
Y WINDSHIELD
Z LEFT HAND REAR VIEW MIRROR
AA FORWARD BILGE PUMP DISCHARGE

Figure 4. Location of various parts of vehicle.

10

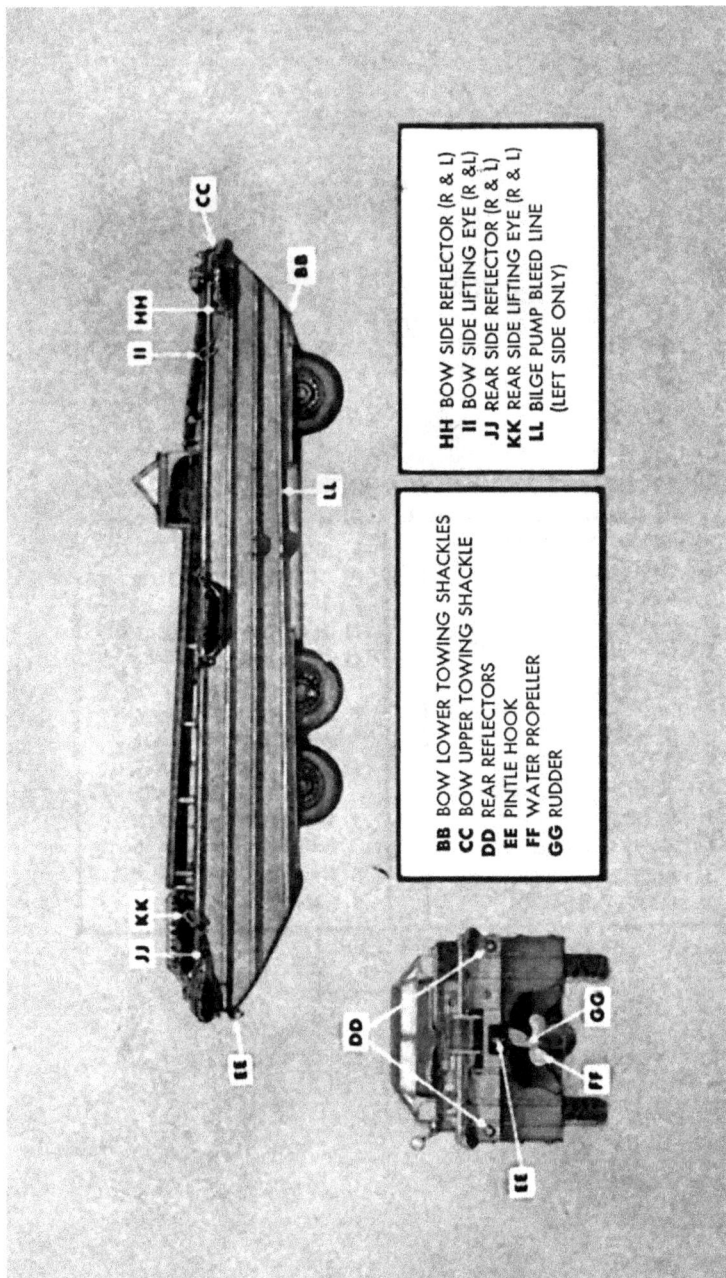

BB BOW LOWER TOWING SHACKLES
CC BOW UPPER TOWING SHACKLE
DD REAR REFLECTORS
EE PINTLE HOOK
FF WATER PROPELLER
GG RUDDER

HH BOW SIDE REFLECTOR (R & L)
II BOW SIDE LIFTING EYE (R &L)
JJ REAR SIDE REFLECTOR (R & L)
KK REAR SIDE LIFTING EYE (R & L)
LL BILGE PUMP BLEED LINE
   (LEFT SIDE ONLY)

Figure 5.   Location of various parts of vehicle.

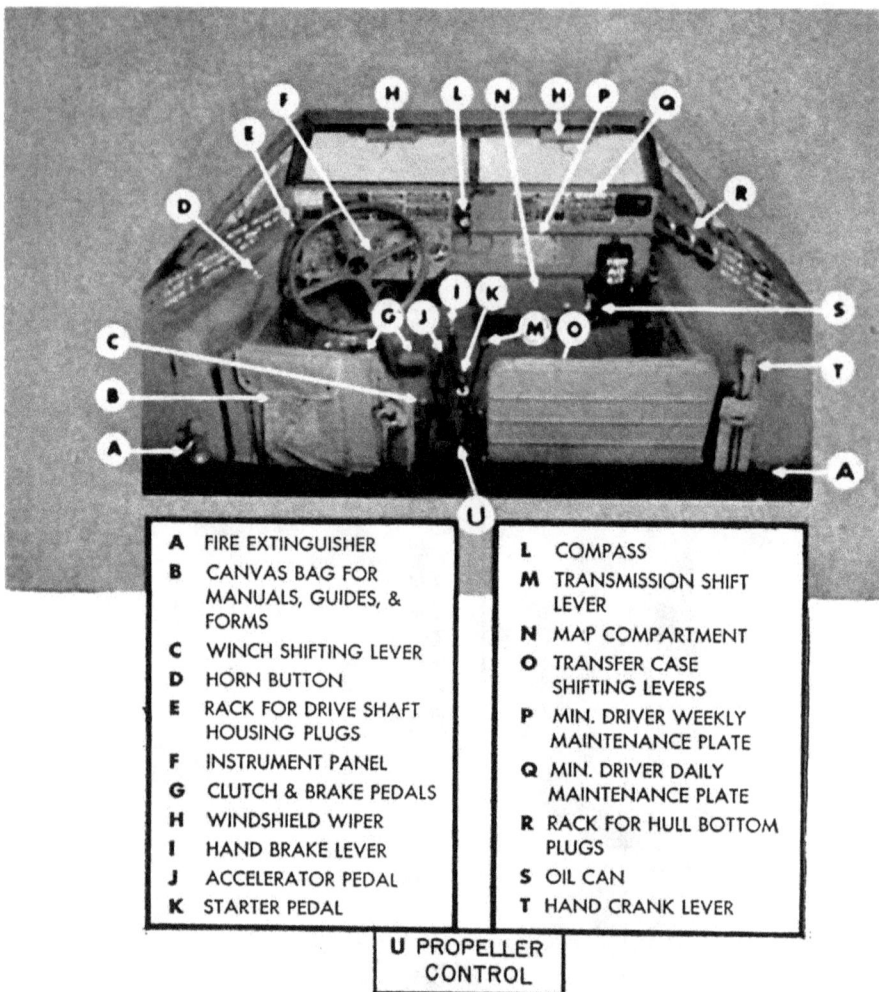

A	FIRE EXTINGUISHER	**L**	COMPASS
B	CANVAS BAG FOR MANUALS, GUIDES, & FORMS	**M**	TRANSMISSION SHIFT LEVER
C	WINCH SHIFTING LEVER	**N**	MAP COMPARTMENT
D	HORN BUTTON	**O**	TRANSFER CASE SHIFTING LEVERS
E	RACK FOR DRIVE SHAFT HOUSING PLUGS	**P**	MIN. DRIVER WEEKLY MAINTENANCE PLATE
F	INSTRUMENT PANEL	**Q**	MIN. DRIVER DAILY MAINTENANCE PLATE
G	CLUTCH & BRAKE PEDALS	**R**	RACK FOR HULL BOTTOM PLUGS
H	WINDSHIELD WIPER		
I	HAND BRAKE LEVER	**S**	OIL CAN
J	ACCELERATOR PEDAL	**T**	HAND CRANK LEVER
K	STARTER PEDAL		

U PROPELLER CONTROL

Figure 6. Driver's compartment.

10. SPECIFICATIONS. a. A comparison of the specifications of a 6 x 6, 2½-ton truck and a DUKW is shown below:

	Regular 2½-ton 6 x 6 truck (Model CCKW–353)	Amphibian 6 x 6 truck (Model DUKW–353)
Over-all length..................	270 inches	372 inches
Over-all height..................	108 inches	106 inches
Over-all width....................	88 inches	99 inches
Ground clearance at hull....		17¼ inches
Ground clearance at front axle...............................	11¼ inches	11¼ inches
Weight, equipped.............	10,300 pounds	15,000 pounds
Total weight including 5,000 pounds payload and driver	15,300 pounds	20,000 pounds
Minimum turning circle diameter.........................	68½ feet	68½ feet
Maximum grade ability (ascending or descending)...............................	60 percent	60 percent
Maximum land speed........	45 mph	50 mph

b. Additional specifications of the DUKW are:

Loaded water line length.............................	344 inches
Loaded freeboard—to deck-front..................	24 inches
Loaded freeboard—to deck-rear....................	16 inches
Loaded freeboard—to coaming front and rear	28 inches
Loaded draft—to front wheels.....................	42 inches
Loaded draft—to rear wheels......................	51 inches
Cargo space—to top of coaming (29 inches deep front, 27 inches deep rear, 82 inches wide, 149 inches long).............................	198 cubic feet
Cargo space under tarpaulin bows................	385 cubic feet
Center of bows above floor, front................	58 inches
Minimum turning circle diameter in water..	40 feet
Maximum speed in water (unloaded)........	6.5 mph
Maximum speed in water (loaded)...........	6 mph

11. CONSUMPTION OF GASOLINE AND SPEED. The economy range, that is, the distance in miles covered on water per gallon of gasoline consumed, in third gear is approximately twice that of second gear, as shown below:

	Gasoline consumption	Speed
Maximum speed in smooth water (loaded), 2d gear (full throttle)..	0.8 mp gal.	6 mph
Maximum speed in smooth water (loaded), 3d gear (full throttle)..	1.4 mp gal.	5.4 mph
Maximum speed in smooth water (loaded), 3d gear ($\frac{1}{4}$ throttle)..	2.3 mp gal.	4.2 mph

SECTION III

SAFETY PRECAUTIONS

---

12. FIRE CAUTION. The DUKW hull is a watertight unit. Gasoline or oil which leaks, or which may be spilled inside the hull, becomes a serious fire hazard and must be cleaned up promptly and thoroughly. The hatches should be opened and the floor boards removed to allow all possible ventilation. The advice given on the various warning and instruction plates must be followed. The fire extinguishers must be kept full and in good working condition. Smoking is never permitted on the DUKW.

13. CLEAN BILGES. Bilges must be kept clean. Rubbish, dirt, or other material dropped in the DUKW eventually will get into the bottom of the hull unless it is cleaned up promptly. If allowed to remain in the hull, it is likely to clog pump intakes, drain valves, radiator core, and other places where it may cause damage serious enough to sink the DUKW. Sand is constantly carried aboard by personnel or with cargo and must be cleaned out immediately to insure proper operation of the pumps.

14. HULL LEAKS. When bilges are cleaned, particularly when afloat, the hull should be checked for leaks. If any leaks are found, their location should be noted and reported so that the maintenance crew may repair them as soon as practicable. As soon as the DUKW enters the

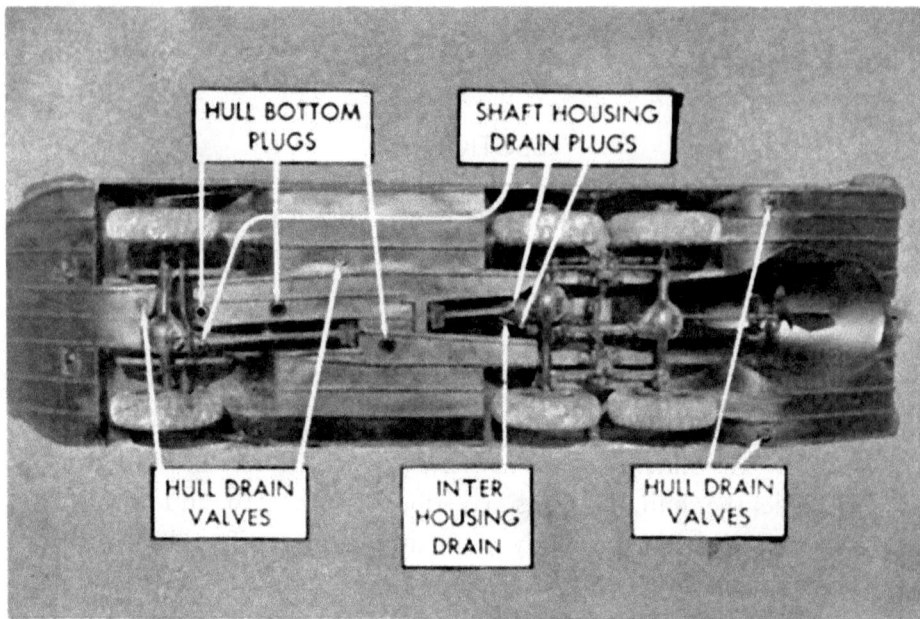

HULL BOTTOM PLUGS

SHAFT HOUSING DRAIN PLUGS

HULL DRAIN VALVES

INTER HOUSING DRAIN

HULL DRAIN VALVES

Figure 7.   Be sure all plugs and valves are tight before entering the water.

water, the driver should look down through the air-intake grilles.   If there is an abnormal quantity of water in the bilges, he should return to land immediately.

15. DRAIN VALVES AND PLUGS. All DUKW's are fitted with three, 3-inch diameter hull bottom plugs and three, 1½-inch diameter shaft housing drain plugs.   (See fig. 7.)   In addition, all DUKW's whose serial number is above 353-2006 have four hull drain valves.   Before DUKW's are driven into the water, a careful check must be made to insure that all plugs are properly installed.

a. At no time, except in storage, will the hull and shaft housing plugs be left out.   They may be removed only temporarily for draining oil or cleaning center bilge.   As a precaution, any plugs removed must be attached to the steering wheel or placed in some other conspicuous place in the driver's compartment to remind him to install them before operating the vehicle. In vehicles not fitted with hull drain valves (serial numbers below 353–2006), only the plug below the transfer case may be left out to drain the hull. When removed, the plug will be attached to the steering wheel.

   b. The three 1½-inch diameter shaft housing drain plugs must be removed and replaced for weekly lubrication of **U**-joints. They also must be removed to drain housing after operating, or immediately on leaving water when the temperature is below freezing. Plugs must be replaced immediately after draining except when the vehicle is to be stored or shipped, or when parked in weather below freezing. In DUKW's which have a serial number above 353–3620, the upper rear housing drains automatically through the flexible tube to the lower housing. The lower front and rear drive shaft housing plugs need only to be removed for draining shaft housings.

   c. When the DUKW leaves the water, the four hull drain valves should be opened to drain the hull. As soon as the hull is drained, the four valves should be closed. All four valves should be kept in open position when the vehicle is being stored or shipped. In the event of valve failure, the hole in the drain valves is tapped to permit the emergency use of the small screw plugs, which are located in the rack on the left side of the engine space. The defective valve should be repaired or replaced as soon as possible.

   Note. On DUKW's with serial numbers from 353–2006 to 353–4201, the forward compartment valve handle is in the engine compartment. On serial numbers 353–4202 and up, the valve handle is located in the left forward corner of the driver's compartment.

16. TARPAULINS. When DUKW's are operated in rough water or heavy surf, and particularly when they are carrying heavy loads, proper use of the cargo tarpaulins is important. Decks and deck openings should be kept watertight, not only to reduce leaking, but also to prevent damage to the engine and to other machinery and equipment in the hull by salt water. When work is done on the engine in rough water, it will often be necessary to use the cab top or other suitable material as a shield to keep water and spray out of the engine hatch and off the engine. See TM 9–802 for proper rigging of tarpaulins.

17. OVERHEATING. The cooling system of a DUKW differs fundamentally from that of truck. The motion of the DUKW does not force air through the radiator, as in the case of the truck. (See fig. 8.) The engine fan circulates all the cooling air through the radiator. To keep the engine cool in hot weather, all parts of the cooling system must be properly maintained. When the cooling

Figure 8. Engine-cooling system showing air passages.

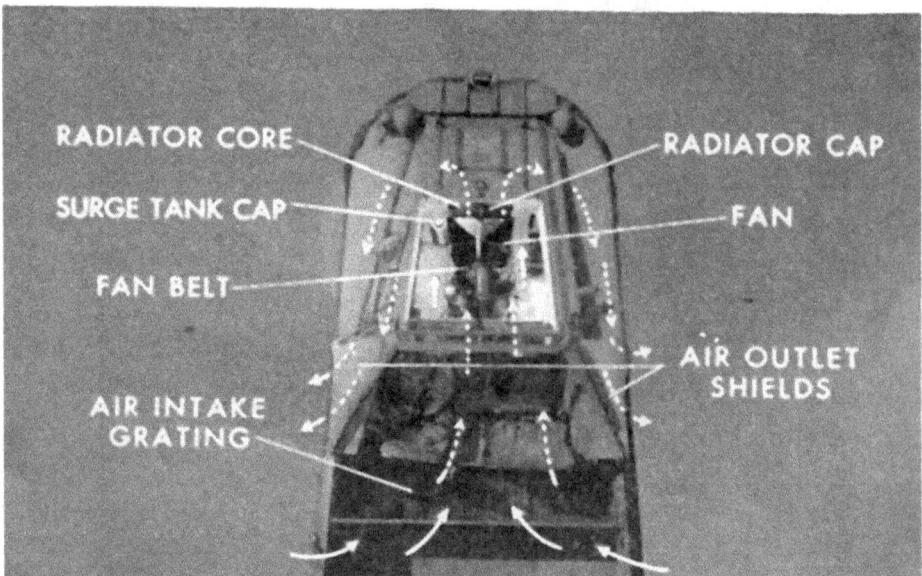

system is working properly, it is not necessary to open the auxiliary air intake or bow hatch. When these are open they admit salt water which may damage many parts of the engine, exhaust, and electrical system. To insure cool engines at all times personnel will—

a. Maintain proper fan belt adjustment.

b. Keep radiator filled with clean water.

c. Keep surge tank half filled where no antifreeze is used.

d. Keep air intake passages clear by carrying no cargo, equipment, or personnel on intake grating, keeping space below grating clear and by stowing tarpaulin bows in bow compartment and not below intake grating.

e. Keep air outlets clear and outlet shutters latched fully opened.

18. TURNING OFF IGNITION. Before the ignition is turned off, the engine should be allowed to idle slowly as provided in TM 9–802. This will prevent the motor from continuing to run, and will prevent damage from sudden cooling to the type of exhaust manifold on early models of the DUKW.

19. LIFE PRESERVERS. All personnel aboard a DUKW will wear life preservers when afloat.

20. MOVEMENT ON DECK. As a general safety measure, especially at night, personnel should not walk outside the cargo space on the decks. However, when it is necessary to move on deck, personnel should always maintain a firm hold with at least one hand.

21. MAN OVERBOARD. Should a man fall overboard, the driver will immediately put the wheel hard over toward the side from which the man has fallen in order to swing the stern and propeller away from him. A life ring or other object that floats should be thrown overboard to mark the spot where the man is located. All hands must keep their eyes on the man in the water since, in a rolling sea, particularly at night, small floating objects quickly move out of sight. The DUKW should be slowed down and, if necessary, the engine stopped so that the man may be located by the sound of his outcries.

SECTION IV

OPERATIONAL MAINTENANCE

---

22. ASSIGNMENT OF VEHICLES. In order to assure proper care of vehicles and equipment, each DUKW should be assigned to a driver and assistant driver who will be held responsible for them. Furthermore, commanding officers should stimulate incentive to maintain mechanical condition and general appearance of equipment. This may be done by the naming of vehicles, and by stenciling drivers' names on windshields, etc.

23. SCHEDULED PREVENTIVE MAINTENANCE. **a.** Because of the DUKW's unusual construction and the nature of its functions, it becomes exceedingly important that preventive maintenance be carefully and completely accomplished. Preventive maintenance duties are described in detail in TM 9–802. Carelessness in following these instructions, or the lack of regularly scheduled preventive maintenance, will result in overloading maintenance personnel and the operation of all vehicles of the company will be affected.

b. Regularly scheduled preventive maintenance also permits the driver and maintenance personnel to observe the condition of the DUKW and to prevent mechanical difficulties. Normal preventive maintenance services are divided into four types: daily, weekly, monthly, and semiannual. When DUKW operations have been unusually heavy, additional preventive maintenance will be necessary.

SUGGESTED SCHEDULE FOR PREVENTIVE MAINTENANCE

	MON	TUE	WED	THURS	FRI	SAT	SUN
6 MONTH PREVENTIVE MAINTENANCE (REPEAT EVERY 26 WEEKS)	1 →	2 →	3 →	4 →	5 →	6 →	7 ... 8 ... 9
1 MONTH PREVENTIVE MAINTENANCE (REPEAT EVERY 4 WEEKS)	8 9 10 11 12 13 14	15 16 17 18 19 20 21	22 23 24 25 26 27 28	29 30 31 32 33 34 35	36 37 38 39 40 41 42	43 44 45 46 47 48 49	50
WEEKLY PREVENTIVE MAINTENANCE	1 2 3 4 5 6 7	8 9 10 11 12 13 14	15 16 17 18 19 20 21	22 23 24 25 26 27 28	29 30 31 32 33 34 35	36 37 38 39 40 41 42	43 44 45 46 47 48 49 50
AVAILABLE DUKW EACH DAY	43 41	41	41	41	41	42	42

Note: The schedule is displayed across roughly four repeating weekly cycles (MON–SUN). The weekly preventive maintenance pattern (DUKWs 1–50 spread across the week) repeats each week.

24. SUGGESTED PREVENTIVE MAINTENANCE SCHED-
ULE. In order to accomplish scheduled preventive maintenance it is necessary that the DUKW's in each company
be numbered consecutively from 1 to 50. These numbers,
together with the days of the week, should be stenciled on
the DUKW as follows:

Stenciled number on DUKW	Days of week stenciled on DUKW
1 to 7	Monday
8 to 14	Tuesday
15 to 21	Wednesday
22 to 28	Thursday
29 to 35	Friday
36 to 42	Saturday
43 to 50	Sunday

The schedule for the semiannual, monthly, and weekly preventive maintenance is shown in figure 9. The four types
of preventive maintenance will be performed as follows:

a. Daily. Accomplished by the driver. Daily preventive maintenance is divided into three operations as follows:

(1) Before operations, or on relieving another driver.

(2) During operations, performed during the time that
the DUKW is not running or is traveling light.

(3) After operations, or before turning over to a relief
driver.

b. Weekly. Accomplished by the driver and assistant
driver with necessary help from maintenance personnel.
Weekly maintenance will be performed on the day of
the week stenciled on the DUKW.

c. Monthly. Accomplished by maintenance personnel
assisted by the driver and assistant driver as necessary.

23

Time not needed for actual mechanical work should be devoted to cleaning and painting the vehicle.

d. **Semiannual.** Accomplished by maintenance personnel assisted by the driver and assistant driver as necessary. Time not needed for actual mechanical work should be devoted to cleaning and painting the vehicle. Semiannual preventive maintenance will normally require that the DUKW be out of operation for a week, although it is possible that it may be completed in 4 or 5 days.

25. SPARE PARTS. When spare parts are not available for DUKW's awaiting repairs, parts may be improvised by making use of demolished equipment, both our own and that captured from the enemy. As a last resort, parts may also be removed from equipment that is beyond repair.

SECTION V

LAND OPERATIONS

---

26. LEVEL ROAD OPERATIONS. a. General. The DUKW has the uses and characteristics of a conventional $2\frac{1}{2}$-ton, 6 x 6 truck, and, on land, is operated in the same manner. It has six driving wheels propelled by a six-cylinder engine through the transmission, transfer case, and shafts. Instead of dual wheels on the rear and intermediate axles, the DUKW is equipped with single wheels on which are mounted 11.00/18–10 ply desert type tires.

b. Speeds. Maximum speeds for each gear are indicated on the caution plate in the driver's compartment. Operators will keep below the maximum speed shown for each gear. Excessive speeds, such as those which occur when coasting downhill in gear or when the governor is not working properly, will damage the engine. In addition, operation at maximum speed leaves no reserve speed to be utilized in emergencies to avoid accidents. The engine will be damaged if allowed to "labor" in gear.

c. Starting. Selection of proper gear combination, together with proper operation of the clutch, is most important when starting the vehicle from a dead stop. When the DUKW is lightly loaded or empty, high-range second gear may be used when starting on level ground, and high-range first gear used when in a hole or on an upgrade. When the DUKW is heavily loaded, high-range first gear may be used on level roads, low-range second with front wheels engaged if off the road, and low-range first with front wheels engaged if in a hole or on an upgrade. When the vehicle is stuck, low-range first or reverse gear, with front wheels engaged, should always be used.

d. Brakes. Whenever brakes are applied, heat is generated and the brake lining is worn. Brakes may be preserved by using the engine to slow down. The driver should learn to slow down well in advance of coming to a stop by releasing pressure on the accelerator while leaving the clutch engaged. The brakes should then be applied firmly but gradually. The clutch should not be disengaged until the truck has almost stopped. It should be remembered that increasing the load increases the stopping distance and that brakes do not hold as well when wet. The hand brake should not be used to stop or slow down except in an emergency. Soon after leaving the water, the brakes should be operated and tested for proper functioning. This is especially important where standard 6 x 6 truck brake lining is being used.

27. OPERATION ON HILLS. a. Uphill. (1) *General.* Before the DUKW is started uphill, the gear which will bring it to the top should be engaged. If the engine stalls on a steep hill, the driver will apply the foot brake and set the hand brake, shift into neutral, and start the engine. Then, using a lower gear, he will make another attempt. If the hill cannot be made, he will shift into reverse and back down the hill. The driver will never step on the starter when the vehicle is rolling backward and the gears are engaged, even if the clutch pedal is down.

(2) *Front wheel drive.* For all steep hills, where first or second gear or low range of transfer case must be used, front-wheel drive should be used.

(3) *Low range of transfer case.* For extremely steep hills and with loaded vehicle, after front wheel drive is engaged, transfer case should be shifted to low range.

b. Downhill. (1) *General.* The engine will be used as a brake when the DUKW is going downhill, using the same gear that would be used in driving up the same hill. On grades over 30 percent, the lowest gear should be used. Drivers will not go downhill too fast in low gear, since this forces the engine to turn over too rapidly. Burned out bearings may result. The speed will be kept down to or below the maximum speed shown on dash plate for the gear being used.

(2) *Braking.* Brakes will be applied at intervals, only as needed.

(3) *Coasting.* Coasting downhill with the clutch depressed or with the transmission in neutral is dangerous and is prohibited. Drivers will not attempt to change gears after starting downhill. A vehicle gains momentum rapidly on down grades, and when the transmission is in neutral, the speed may be too great to mesh the gears.

28. CONVOY OPERATIONS. a. Command control.

The column commander will control the speed, distance between vehicles, routes, traffic precautions, and other details, with the assistance of his officers and noncommissioned officers. The leading vehicle should be a DUKW. More maneuverable vehicles should bring up the rear. Driver and assistant driver will observe for enemy aircraft at all times.

b. Right-of-way. (1) In a small motor column, each vehicle will extend the same right-of-way as do individual vehicles, unless the column is accompanied by police escort to block off traffic.

(2) In a large motor column, guides may be dropped

from a vehicle which precedes the column. The guides will be picked up by a following vehicle and, at the first appropriate halt, will go to the head of the column so that the same procedure may be repeated.

c. Safe distances in column. Under normal road and weather conditions, the distance in yards between vehicles in a motor convoy will be twice the driving speed. For example, at 25 miles per hour, there should be a distance of 50 yards between vehicles. Under actual or simulated battle conditions, no vehicle will be within 75 yards of any other vehicle. Often on straight level stretches of road it is necessary to keep 400 to 500 yards distance between vehicles as a precaution against strafing by enemy aircraft. While operating on roads with sudden steep grades, it is necessary to maintain sufficient distance to allow vehicles to make the grade without causing congestion at the rear of the column.

d. Blackout lamps. (1) In convoy driving at night, under blackout conditions, the blackout marker lamps are of great assistance in estimating the distances between vehicles.

(2) There are two rear lamps which are divided into two parts, making four parts in all. Under conditions of normal visibility when the vehicle ahead is less than 60 feet away, four lights appear to be visible; between 60 and 180 feet, two lights are visible; between 180 and 800 feet one light is visible; beyond 800 feet no lights are visible.

(3) Under conditions of normal visibility the front marker lamps are visible as two up to 60 feet; from 60 to 800 feet they appear as one; beyond 800 feet they are invisible.

29. SAND OPERATION. a. General. Because of its special type of tires, the DUKW has exceptional ability to operate in difficult sand conditions. To avoid difficulty, the driver must be familiar with and be guided by special points which are covered in the following paragraphs.

b. Front wheel drive. Front wheel drive must be engaged for all operations on sand.

c. Tire pressure. Wherever soft sand is encountered, it is vitally important that correct tire pressure be used. Stencils and dashboard instruction plates and TM 9–802 give detailed instructions which must be carefully studied and accurately followed.

(1) *Tire deflation.* The most important preparation for operation on soft sand is tire deflation. The area of ground contact of a tire deflated to 12 pounds pressure is approximately four times that of the same tire inflated to 40 pounds. (See fig. 10.) The deflated tire tends to travel over the sand, while the highly inflated tire digs in.

(2) *Changing pressure.* The amount of tire deflation necessary depends on the consistency of the sand. Based on a normal payload of 5,000 pounds, 12-pound tire pressure should be used for *very soft* sand. This pressure should be increased as necessary, going up to 20 pounds for moderately soft sand. Tire pressure should be increased 1

Figure 10.   Tire deflated to 12 pounds travels over soft sand while tire inflated to 40 pounds sinks in.

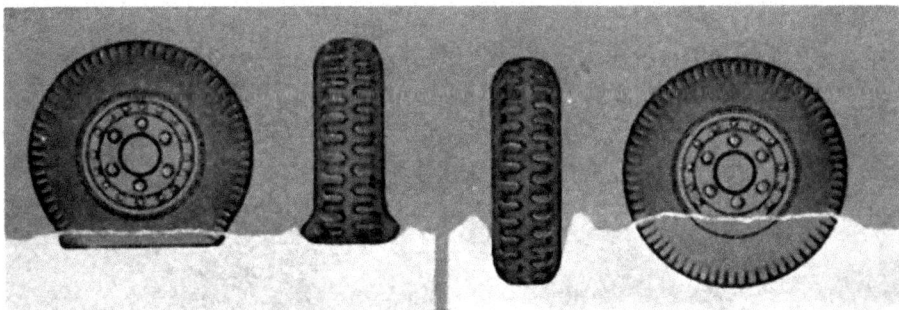

Figure 11 ①. *Cannot get over but can back out.*

pound for each 1,000 pounds overload above the 5,000 pounds rated payload. Thus with 8,000 payload, in very soft sand, the minimum pressure should be 12 pounds plus 3 pounds or 15 pounds. If the truck is to be stored for any length of time after low pressure tire operations, tires should be inflated. This will help to preserve them.

(3) *Central control system.* The DUKW's after serial number 353–2005 have a central tire control system which will inflate or deflate the tires on all wheels at the same time while the vehicle is stopped or in motion. This control can be operated by the driver without requiring him to leave his seat.

d. Avoid spinning wheels. Drivers must not "dig in" but, with tires properly deflated, travel over the sand with

Figure 11 ②. Result of spinning—must be winched out.

full power. It is important to roll fast and keep rolling. If the DUKW stops and wheels are permitted to spin for over a few seconds, digging in has started, the roadway is damaged, and the vehicle may have to be towed or winched out. This must be avoided. When digging in starts, drivers should disengage the clutch immediately. It is then advisable to back out and make another try in a better place, if that is possible. (See fig. 11 ① and 11 ②.)

e. **Follow the leader.** If a group of DUKW's is operating in and out of water, or over a sandy beach, the leader should pick the best available path. Those behind should follow in the leader's tracks as long as they are passable. Drivers will not follow too closely, since the vehicle ahead may get stuck and have to back up to make

31

another try forward. It is advisable to keep twice the normal road distance between vehicles when operating on sand or difficult terrain.

f. Grades. DUKW's will be driven up or down difficult sand grades *straight* and *square*. Drivers must learn to avoid crossing a hill at an angle or straddling the vehicle on the crest of a dune. When traveling over sharply rising or falling sand dunes, where there is a possibility of bottoming, DUKW's will be kept rolling fast. The sky line should be avoided when the enemy is in the same area.

30. DRIVING OVER DIFFICULT TERRAIN. a. General.
Tree stumps, sharp rocks, and other such objects which might puncture the hull will be avoided. Soft mud should always be avoided since no wheeled vehicle can operate in it.

b. Wire. In operations where wire may be encountered, the front brake shut-off cock will be closed. It is located in the right frame side rail below the middle of the driver's compartment. With the shut-off cock closed, the braking system will not become inoperative if wire damages the front brake hose.

c. Land under shallow water. When driving on land or along the beach, the operator should also avoid areas covered with shallow water. Water-covered ground may contain deep holes, wreckage, or obstructions that may cause damage to the tires, chassis, or hull of the vehicle.

SECTION VI

WINCHING OPERATIONS

---

31. GENERAL. **a. Uses of winch.** Wherever the DUKW is operated, frequent use is made of its winching equipment. A thorough knowledge of winching techniques, circumstances in which the winch may be used, and methods of rigging are important. The winch, because of its great power, can be of assistance under many conditions, and, as the drivers become familiar with it, they will learn the many uses to which it may be put. It may enable the DUKW to extract itself after being stuck. It will frequently be used to assist other vehicles and landing craft. It will be used with the **A**-frame for loading and unloading.

b. How used. The winch on the DUKW is basically the same as that on a regular 6 x 6, 2½-ton truck except that on the DUKW it is mounted at the stern. (See fig. 5.) In this position it can be used for pulling either to the front or back. On the DUKW, when the winch is to be used forward, the cable must be led through the winch guides in the rear coaming, in the base of the windshield, and through the bow fairlead. When winching astern, the cable leads directly from the winch drum.

c. Towing. The winch cable *must not* be used for land towing. This is likely to jam and kink the cable on the drum. Instead, the tow chain is used from the stern pintle hook to one of the low bow shackles on the other DUKW. (See fig. 12.)

33

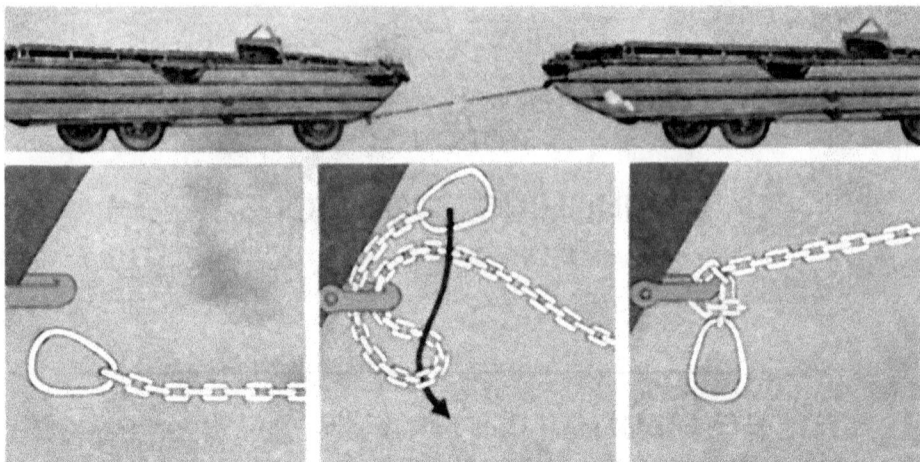

Figure 12. Rigging tow chain from pintle hook to one of the low bow shackles. Note detail for securing chain to shackle.

32. DETAILS OF WINCH. a. Cable and shear pins.

The winch is equipped with 150 feet of half-inch diameter galvanized cable with breaking strength of approximately 18,000 pounds. The cable is protected by a shear pin in the winch drive mechanism which will break when the load on the cable exceeds approximately 10,000 pounds. The shear pin is located below the rear deck at the coupling where the winch drive shaft goes through the stern. Spare shear pins are kept in a rack close to this position.

b. Power take-off lever. The winch is operated by a power take-off on the left side of the transmission. A lever coming through the cab floor enables the driver to operate the winch in high, low, and reverse gears with two neutral positions in between. A hinged locking plate attached to the cab floor provides a positive means of holding the lever in the neutral position between reverse and low. This is to prevent accidental engaging of the power take-off. An automatic safety brake on the worm shaft will support a load on the winch cable when the lever is being shifted into various operating positions.

c. Sliding jaw clutch. The sliding jaw clutch on the rear deck is used to engage and disengage the winch drum. (See fig. 13.) This sliding jaw clutch must be engaged for the winch power drive to have effect. When in the disengaged position (handle moved toward the left side of the DUKW), the winch drum will rotate freely on its shaft.

d. Drum flange lock. The drum flange lock is mounted on the sliding jaw clutch handle mechanism. (See fig. 13.) This lock is spring loaded and can be made to engage one of several holes in the winch drum flange when the sliding

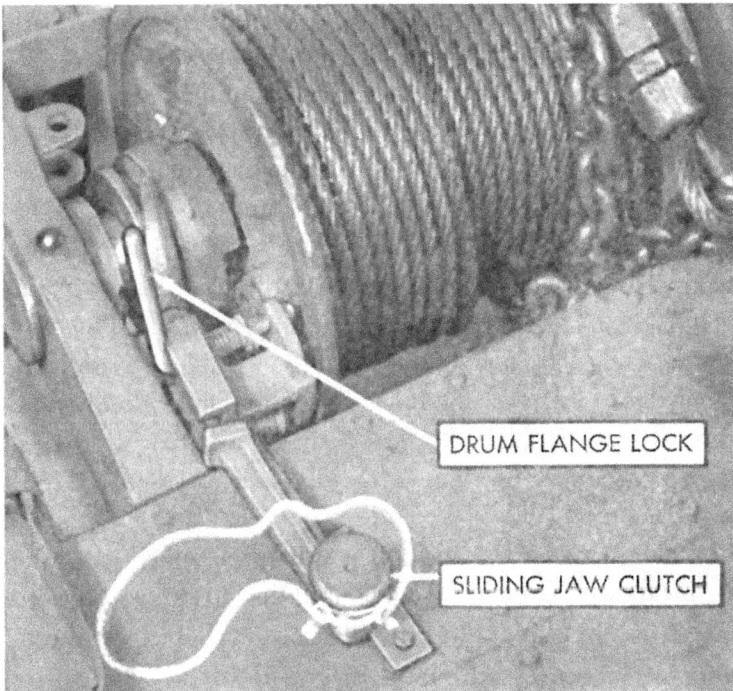

DRUM FLANGE LOCK

SLIDING JAW CLUTCH

Figure 13. The winch is located on the rear deck.

jaw clutch is in released position. The function of this pin is to prevent damage if the power take-off lever is accidentally shifted and also to prevent the cable from unwinding while the vehicle is being operated, the jaw clutch being normally left in the disengaged position.

33. OPERATING CYCLE. a. To hook up. To hook up the cable, the sliding jaw clutch lever is placed in disengaged position and the drum flange lock handle is disengaged and swung up. As much cable as is needed to make the desired hook up is pulled out.

b. To wind in. The sliding jaw clutch lever is placed in the engaged position and the winch is operated by shifting the power take-off lever in the cab, using the clutch in the same way as when driving. Normally, high speed should be used to take up slack in a loose cable or for extremely light loads, and low speed should be used whenever heavy pulling is to be done. Reverse is used to pay out cable. Reverse is not used for winching to the rear, since either low or high gear will wind in the cable regardless of the direction in which the cable is led. Before the operator starts to wind in the cable, the cable already on the drum must be tightly and evenly wound. This will provide a smooth surface for additional cable to wind on. The assistant driver should watch the cable closely and guide it, if possible, with the hand crank bar so that it winds evenly and does not become kinked or jammed. (See fig. 14.)

c. Engine operating speeds. When the winch is in use, the maximum engine speed should be 1,000 revolutions per minute or about one-third throttle. The winch mechanism and bearing will be damaged if higher speeds are used.

d. To stop. To stop the winch, the driver should

Figure 14. Crank bar is inserted in pintle hook to guide cable.

depress the clutch and shift the power take-off to neutral.

e. To pay out cable by power. When cable is under load, the winch power take-off must be shifted into reverse and the winch cable payed out by power until it becomes slack. When the cable is slack and there is more than about 10 feet of additional cable to be payed out, the jaw clutch should be disengaged and the cable pulled out by hand. The paying out of cable should be stopped when at least five turns remain on the drum or the cable load will fall directly on the end attachment and the cable is likely to pull out. When another vehicle is pulling out the cable, the same procedure applies as when the cable is under load. When afloat it is also important to stop the cable from paying out completely. As a last resort, the cable can be stopped by shifting the jaw clutch to the engaged position.

f. When not in use. When the winch is not in use, the cable will be cleaned and well lubricated to prevent rust, and wound up neatly on the drum. The power take-off will be locked in neutral, the winch jaw clutch shifted to disengaged position, and the flange locking pin engaged to prevent cable unwinding.

34. RIGGING METHODS. a. General. In winching operations, an adequate rigging hook-up made at the start will eliminate failures which waste time, damage equipment, and often leave a vehicle more seriously stuck than before.

b. Winching points. The term "winching point" is applied to any fixed object to which the winch cable or rigging is attached during winching operations. The selection of a suitable winching point is very important. Its position with reference to the vehicle that is stuck and also its ability to withstand the necessary pull are vital to successful winching operations. It is desirable to locate the winching point as nearly as possible in line with the direction in which the stuck vehicle is to be moved. It should be sufficiently far away so that most of the cable is unwound from the drum in making the necessary hook-up. However, at least five coils should remain on the drum. The question of single, double, or triple rigging will be determined by the distance of the winching point.

c. **Types of winching points.** Winching points may be any of the following:

(1) *Other vehicles.* The most practical winching point is one or more vehicles, such as DUKW's, trucks, tractors, etc., since they can be moved to exactly the right position.

(2) *Trees.* If no suitable vehicle is on hand, a substantial, well-rooted tree is the next best choice.

(3) *Anchor.* Where no trees are available, the anchor carried on each DUKW will provide a satisfactory winching point for moderate pulls. The anchor is particularly effective in soft sand where it will dig itself down to about 3 feet below the surface. In this position it will afford a pull of approximately 7,000 pounds.

(*a*) In using the anchor it should be placed so that there is no sideways pull or stress on the anchor bar, and at least 75 feet of cable should be used to prevent too great an upward pull. The anchor may drag 15 to 30 feet before it digs down far enough to hold. It will not provide a sufficiently firm winching point to make more than a single rigging effective.

(*b*) Where more pull is needed, a second anchor may be used. The second anchor should be hooked in tandem to the first anchor by using the tow chain or auxiliary cable. The two anchors must be some distance apart and will allow as much as 14,000 pounds of pull. In this case, a double rigging will be more effective.

(c) If there is vegetation on the sand which prevents the anchor from digging deep enough to hold, or if the ground is hard, the anchor should be buried in a manner similar to that described in (4) below for the deadman. (See fig. 15.)

(4) *Deadman.* When no suitable vehicle or tree is available and when the anchor does not prove effective, a deadman can be used. Although this takes considerable time to set up, if properly installed it can withstand a heavy pull. (See fig. 15.)

d. Details of cable hook-ups. The effective power of the winch can be varied by a number of cable hook-ups, involving the use of one or more snatch blocks and, on occasion, one or more auxiliary cables. Each DUKW is equipped with one snatch block and one 75-foot auxiliary cable. When unusual winching problems arise, it may be

Figure 15. Rigging a "deadman."

necessary to borrow additional cable and snatch blocks from
other DUKW's. The power of any given hook-up will be
greatest when the cable is paid out to the bottom layer
on the drum. This increases the power by reducing the
effective drum diameter and produces the maximum of
about 10,000 pounds tension on the cable. A full drum
will produce only about half this cable tension before snap-
ping the shear pin. The following methods should be
used in rigging the cable:

(1) *Single rigging.* The simplest hook-up is a direct
cable lead forward or backward and is called a single
rigging. (See fig. 16.)

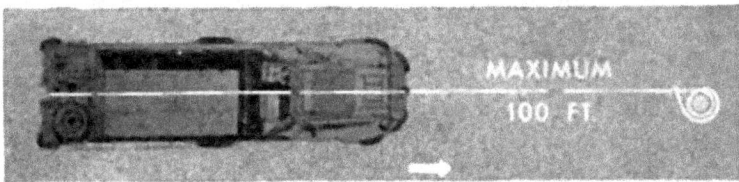

Figure 16. Single rigging.

It gives a maximum of 10,000 pounds pull (capacity of
shear pin) and is desirable because of its simplicity and
also because it provides the maximum possible reach from
the winch. This reach can be increased if necessary by
using auxiliary cables, tow chains, ropes, or any other
usable extensions.

(2) *Double rigging.* When the 10,000 pounds of pull
provided by single rigging is not sufficient, the pulling
power of the winch can be doubled. (See fig. 17.) This
can be done by leading the cable through a snatch block
which is secured to the winching point, then bringing
the end of the cable back and hooking it to the object
to be moved. This will produce a minimum effective pull

41

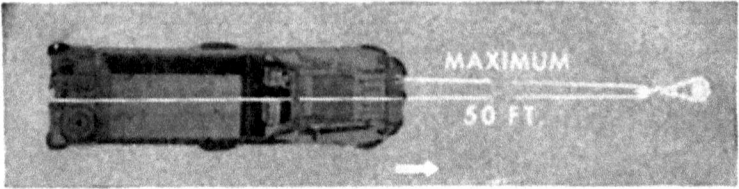

Figure 17. Double rigging.

of 20,000 pounds. The range, however, is reduced by the doubling of the cables. The snatch block normally is attached to the winching point by use of the tow chain. If the double hook-up is necessary for power but there is not sufficient reach, the auxiliary cable can be hooked between the snatch block and the winching point, extending the range 75 feet, if it is single, or 37½ feet, if looped around double.

(3) *Triple rigging.* Additional winching power can be secured by using two snatch blocks and tripling the cable. (See fig. 18.) This will provide up to 30,000 pounds effective pull. The triple rig is made by leading the cable through a snatch block at the winching point, back to a second snatch block hooked on the DUKW and then by securing the end of the cable to the winching point or object to be moved. The range of the triple hook-up is

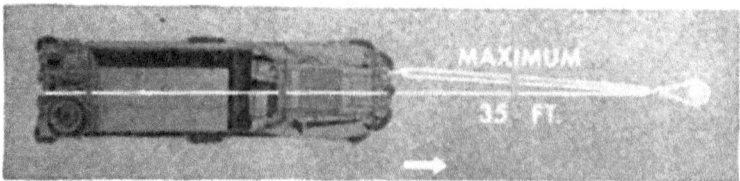

Figure 18. Triple rigging.

necessarily short but can be extended by the use of one or more auxiliary cables and tow chains. Because of the power available, auxiliary cables should always be doubled when used in conjunction with the triple hook-up.

35. DUKW WINCHING ITSELF. The most frequent use of the winch is to enable a DUKW to pull itself out when it has become stuck. The following steps should be taken:

a. Winching point. The driver should first select a suitable winching point, considering its position, distance, and effectiveness in relation to the estimated pull necessary to free the DUKW.

b. Rigging method. He next will select a rigging method consistent with the power necessary to free the DUKW and consistent with the distance to the selected winching point.

c. Procedure. The wheels should not be engaged until the winch cable becomes taut; then the winch should be assisted by engaging the wheels (low low or low reverse). Otherwise, the shear pin may be overloaded and snap or the winching point pulled loose without freeing the DUKW. Usually, it will be necessary to winch only a short distance, but the driver must be sure that the vehicle is in the clear before unrigging the cable.

36. DUKW WINCHING ANOTHER VEHICLE. On many occasions it will be necessary to use the winch to assist another vehicle that is stuck. Even if the stuck vehicle is equipped with a winch, it is better to use the winch on the assisting vehicle. The following steps should be taken:

a. Position. The assisting DUKW should be placed in the best position, considering first that it must not get stuck itself and second, the most practical direction to pull the stuck vehicle.

b. Rigging. First, the driver must select a rigging method consistent with the power necessary to pull out the stuck vehicle.

c. Winch. Then, the winch on the vehicle that is *not* stuck is used so that the stuck vehicle can use its wheels without interfering with the steady pull of the winch cable.

d. Procedure. If the winch pulls the assisting DUKW toward the stuck vehicle, even though hand and foot brakes are set, the winch should be disengaged. The assisting DUKW will then dig in its wheels about a foot by attempting to tow the stuck vehicle. Next the wheels are disengaged, the brakes set, and the winch is tried again. The holes dug by the assisting DUKW, plus hard pressure on the hand and foot brakes, should provide enough anchorage to pull the stuck vehicle free. It may be necessary to secure the assisting DUKW to another vehicle or other winching point to hold it in position. Occasionally two or three DUKW's may have to be used to move one that is badly stuck. Each should rig its own winch cable, each should dig in as necessary, then, all pull evenly.

e. Pulling out. When it has been necessary for an assisting DUKW to dig in to provide necessary pull, it must be free itself before unrigging. This can be done by having the assisted DUKW hold its brakes to serve as a winching point. The assisting DUKW may then, by means of its own winch, pull out of the hole it has dug.

37. MISCELLANEOUS WINCHING. a. Winching landing craft. Details of this operation are given in paragraph 115.

b. Lifting with A-frame. Details of the operation are given in paragraph 92.

SECTION VII

WATER OPERATIONS

---

38. GENERAL. The DUKW is as much a seagoing craft as it is a land truck. While drivers may be familiar with the land operation of a 2½-ton, 6 x 6 truck, boat handling will be an entirely new experience for most of them. Instructions applicable to water operations, therefore, should be studied carefully. Definitions of nautical terms used in this manual and employed generally when operating on water are listed in appendix III.

39. PREPARING TO ENTER WATER FROM SHORE.
a. Conditions. The principal considerations are: (1) how *steep* is the beach; (2) how *soft* is the sand (can it be crossed and at what tire pressure) ; and (3) how *rough* is the ground (can be seen above water, and must be *estimated* below water). Obstacles under water will be more difficult to negotiate than if they are on dry land. Mud, swamps, marshes, quicksand, stumps, wreckage, large sharp rocks and boulders should be avoided.

b. Tire deflation. A very soft, sandy beach will require tire deflation to a minimum of 12 pounds, soft sand 20 pounds, while reefs of coral or sharp rocks require tire inflation to 30 pounds. See figure 10 and dash instruction plate. For overloads, all recommended pressures should be increased 1 pound for each 1,000 pounds over the 5,000 pounds payload.

45

c. Hatches. All hatch covers must be clamped tightly. Early model vehicles were equipped with auxiliary air intakes which must also be tightly clamped.

d. Hull drain valves, housing drain plugs, and bottom plugs. Before entering water, bilges must be reasonably dry, all four drain valves closed, and all plugs clean and tight. (See paragraph 15 for details.)

e. Tarpaulins. For surf over 5 feet high, cargo tarpaulins will be rigged securely. Tarpaulins must also be used when long hauls are to be made or when the hauls will be made in rough water.

f. Surf. Unless water is smooth, the bow surf plate must be set up. For surf over 5 feet high, the windshield surf guard will be set up on those vehicles with vertical windshields. If the cargo comes up to or above the coamings, the tarpaulin may be lashed directly over the cargo and coamings without the use of the tarpaulin bows. On early model vehicles the wood bows should be reinforced with the two auxiliary ridge poles.

40. ENTERING WATER. **a. Bilge pumps.** Before the DUKW enters the water, the forward bilge pump is set to drain the center or main compartment. On DUKWS that are fitted with them, the other forward bilge pump valves and the drains are closed. (See dash instruction plate and paragraph 41b below, for pump system details.)

b. Selection of approach. The beach must be surveyed and the most desirable approach selected. This will vary with the location, but a hard, sandy, moderate grade beach is best. Operational beaches are usually littered close to the edge of the water with wreckage, spilled cargo, floating fuel cans, and occasionally with barbed wire, stakes, coral, stumps, boulders and so forth. These should be avoided as much as possible.

46

c. Entering. With the transmission in second gear, the front wheels are engaged and the transfer case placed in low range. When entering water down steep banks or ramps, drivers must proceed *slowly*. The stern pintle hook is set in horizontal position so that it will not strike obstructions. The bow must always face at right angles to the surf. (See figs. 19 and 20.)

d. Engaging water propeller. Unless there is heavy surf or strong current, the water propeller is not engaged until the DUKW is fully afloat. This will decrease considerably the chances of damaging the propeller. If the propeller is damaged, it is immediately disengaged before the hull is punctured and the DUKW should return to shore, using driving wheels only. If there is a heavy surf or strong current, the water propeller should be engaged before reaching the edge of the water. To do this, the DUKW must be momentarily stopped and then proceed with *full throttle*.

Figure 19. Hit the surf square—use full throttle.

41. IN WATER. a. Disengaging driving wheels. The wheels will not be disengaged until the DUKW is clear of breakers and over all outlying reefs. Only about ½ mile per hour speed is lost by having the wheels and propeller engaged at the same time, and the danger of getting stuck on a sand bar is considerably decreased. When the DUKW is in *deep* water, the wheels are stopped by shifting the transfer case into neutral. The driver will not disengage the front wheel drive.

b. Bilge pumps. The DUKW is equipped with three bilge pumps.

(1) *Rear bilge pump.* The rear bilge pump is a centrifugal pump and operates automatically whenever the propeller is driven ahead. It does not begin to pump until the water is about 5 inches deep in the center compartment. On DUKW's prior to serial number 353–2006, it has a capacity of 160 gallons a minute at 2,500 rpm. After

Figure 20. Enter and leave the water at right angles to the waves.

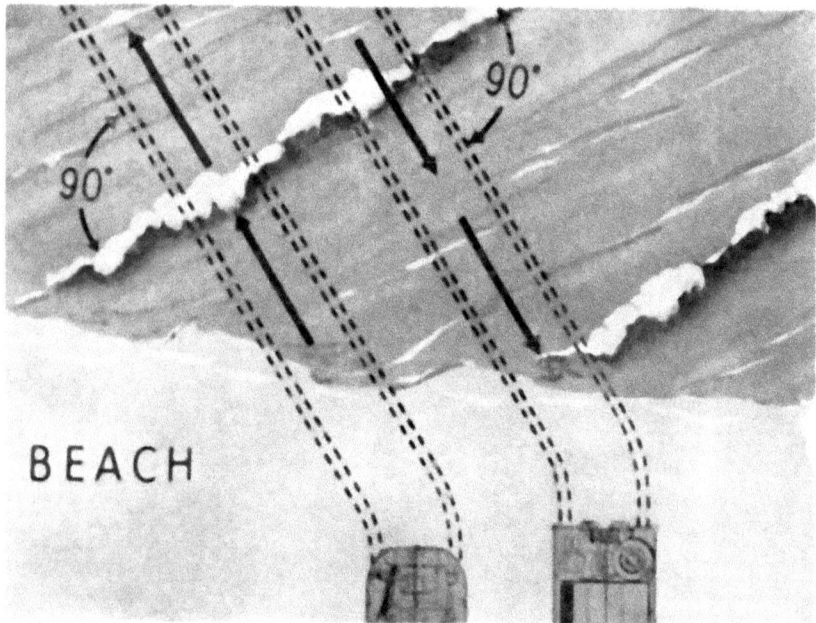

serial number 353–2006, the capacity was increased to 260 gallons per minute.

(2) *Forward bilge pump.* The forward bilge pump below the driver's seat is a gear pump and removes water from the left and the right compartments, from the center compartment and, in some DUKW's, from the bow compartment. The forward pump, will discharge 63 gallons a minute on vehicles below serial number 353–4202. After serial number 353–4202, the pump is a centrifugal pump with approximately the same capacity. It is not practical to pump from more than one compartment at a time. It is also undesirable for the pump to run without having one valve open.

(*a*) On all DUKW's the different valves should be opened alternately, leaving each valve open as long as water continues to come out of the discharge. However, when no bilge water appears to be coming out, it is advisable to operate the different valves in rotation, each valve for a minimum of about 2 minutes. The discharge should be watched. If an abnormal amount of water is being discharged from any one compartment, that section should be inspected for leaks and drain valves should be checked to see that they are shut tightly.

(*b*) DUKW's that have five rods coming up through the floor immediately in front of the driver's compartment have four manifold valves and a manifold drain. The center rod of the five is the manifold drain. Whenever afloat, it is important that this manifold drain be in the closed position (control pushed down). If the control remains *up,* it will interfere with the pumping of the various compartments.

(*c*) DUKW's with serial numbers above 353–4202 have only three controls. These come out through the front face of the driver's seat.

(3) *Hand pump.* In emergencies, when the engine is not running or power pumps are inoperative, the hand pump, buckets, and helmets should be used for bailing. The hand bilge pump will discharge approximately 25 gallons per minute. If necessary cargo may be thrown overboard, particularly at the low side of the DUKW or near the point where the water is coming in. Cargo should be shifted to permit proper use of hand pump, particularly when it is necessary to pump from the rear compartment.

c. Steering. Water steering requires more turning of the wheel than land driving. In water operations there is a considerable lag both at the start and at the finish of a wheel turn. The driver must anticipate turns, particularly in rough water, swinging the wheel ahead of time if possible. In the water, it is the *stern* that responds to the steering wheel, not the *bow*. In addition, the stern momentarily swings *toward* an object as the DUKW is steered *away* from that object. (See figs. 21 ① and ②.) Rudder response will be less at reduced speed. But even at a very low speed, rudder control will be better if occasional short spurts of power are given, causing the propeller to drive sufficient water against the rudder to produce the desired turn.

d. Speed. At 2,500 engine rpm, smooth water speed without payload should be 6.5 miles per hour. This speed will be reduced about 0.1 miles per hour for each 1,000 pounds of load carried, giving 6.0 miles per hour maximum speed with 5,000 pounds of load and 5.7 miles per hour maximum with 8,000 pounds of load. In addition, water speed will be reduced about 0.1 miles per hour for each foot of wave height encountered. Thus a DUKW with 7,000 pounds load in 3 foot waves should make 5.5 land miles per hour (losing 0.2 for overload and 0.3 for wave height). Performance figures are given on dash-

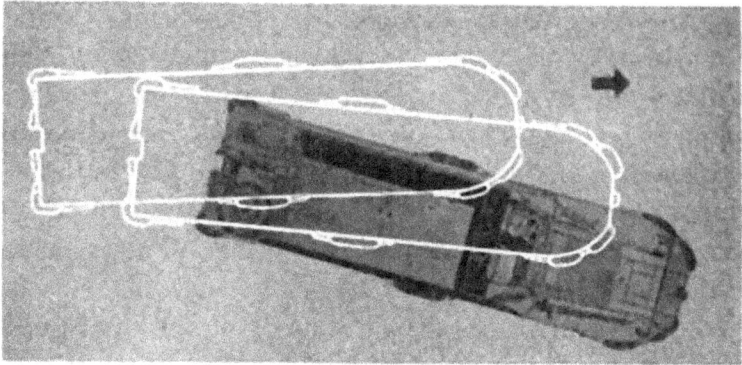

Figure 21 ①. *Steering on land—bow moves
in the direction of the turn.*

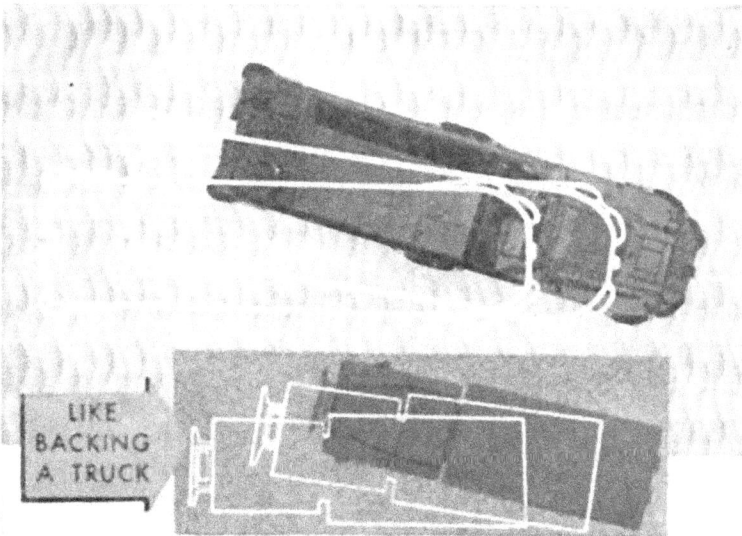

LIKE
BACKING
A TRUCK

Figure 21 ②. *Steering in water—stern moves
opposite to the direction of the turn.*

board plates. These figures are all based on *land* miles. (See paragraph 80 for conversion into nautical miles.)

e. Endurance. For runs where capacity of gasoline supply is not a factor, full throttle, with transmission in second gear high range should be used. The propeller holds the engine down to reasonable operating speed of about 2,500 rpm. On long runs when gasoline must be conserved, engine speed should be dropped to 2,100 to 2,200 rpm in second gear or to full throttle in third gear. The engine will have a tendency to run hot in third gear, but the temperature can be brought down by periodically running the engine in second gear for several minutes. The most economical gear for gasoline conservation is third gear, quarter throttle. Speed, however, is reduced. (See paragraph 11.)

f. Range. The maximum water range of the DUKW, without increasing the fuel supply, is 28 *nautical* miles. This is equal to 32 *land* miles.

g. Manifold glow. At night the exhaust manifold glows cherry red. This is normal, and no sign of trouble.

h. Refueling. Refueling on the water is difficult. The DUKW should be maneuvered so that the gas tank filler neck is facing away from the waves. Rags should be kept handy to block out water as necessary.

i. Reversing. It is impossible to stop the DUKW or any water craft quickly in water; reversing the propeller is the only means of stopping. To do this the driver will shift transmission into reverse, and, on DUKW's with two speed propeller transfer case, shift this to reverse position also. Full throttle is used when reversing, with all wheels engaged in high range to get maximum power. When the DUKW is traveling light, reversing is greatly facilitated by shifting the crew and any movable weights to the rear to keep the propeller tunnel under water.

j. Precautions. The DUKW must be kept clear of lines, moorings, anchor cables, nets and buoys, particularly on the windward or upcurrent side. Loose lines must be kept clear of the propeller, particularly when towing. If the DUKW fouls a line or gets too close to one, drivers should disengage the propeller and wheels (if engaged) at once, and use the boat hook to clear the lines, and drift away. If a line becomes fouled in the propeller, momentarily reversing and then going ahead again will often loosen it. This procedure repeated several times may throw the line clear without causing damage. In no case will the propeller be forced to turn when it is being held by a fouled line.

42. PREPARING TO LAND. a. General. The same careful planning and skill must be used in bringing the DUKW out of the water as is used when entering. The landing point should be reconnoitered, if possible. When in doubt, drivers must prepare for the most difficult conditions.

b. Landing spot. The most favorable available spot for landing should be chosen. The principal considerations are (1) how steep is the beach, (2) how soft is the sand (can it be crossed, and at what tire pressure), and (3) how rough is the ground (can be *seen* above water, and must be *estimated* below water). Obstacles under water will be more difficult to negotiate than if they were on dry land. Steep beaches, mud, swamps, marshes, quicksand, stumps, wreckage, large sharp rocks and boulders should be avoided. If the DUKW is equipped with central tire control, tire pressure should be set to best suit the conditions.

c. Approach. Drivers should engage all driving wheels while still outside the line of surf. When landing conditions are unknown, wheels should be engaged at least 300 yards outside the breakers.

d. Surf. In rough weather the rear surf shield is set up, and, if the surf looks heavy, the tarpaulin rigged. Before entering the breakers the driver should get in a position that will allow him to approach the desired landing point with the waves directly astern. Full throttle is used while going through the breakers. Drivers should be alert to counteract any tendency of the waves to turn the DUKW off its safe course.

e. Current. DUKW drivers must learn to observe the effect of the current. When there is a strong current running, the DUKW should be about 50 to 100 feet upcurrent from the desired landing spot while still about 150 feet out from shore. The DUKW is then turned down current toward the landing spot under full throttle so that when its front wheels ground the DUKW will be close to the desired spot and heading about 30° down current. Immediately after the front wheels have grounded, the current will swing the floating rear of the DUKW down current, straightening it out. The front wheels must then be turned sharply down the current to prevent the stern being carried around beyond the desired landing course.

43. LANDING. a. All landings on unfamiliar beaches should be made at *full throttle* with the transmission in second and the transfer case in low range. This does not apply for coral, when instructions outlined in paragraph 44 should be followed. DUKW's should come in at right angles to the waves. (See fig. 20.) If the waves are at an angle to the shore, drivers should square around slowly after getting safely past the surf and drive straight up the beach, continuing to operate at *full throttle* until well ashore. On a rocky or coral beach, DUKW's must be driven *very slowly*.

b. On beaches that are clear of obstacles, or in heavy

surf or strong current, the propeller should be disengaged as the DUKW leaves the edge of the water. It is not necessary to disengage the clutch or to slow down. Since it is no longer under load, the propeller will disengage easily and quietly as soon as it lifts clear of the water.

c. On a rough beach, on coral, and wherever wreckage may be in the water, it is very important to *disengage the propeller at the first possible moment.* This can be done as soon as some traction can be obtained from the wheels. This will greatly reduce the chances of damage to the propeller.

d. In sand or mud, if forward progress stops because the wheels spin, the clutch must be disengaged *immediately. The wheels must never be allowed to dig in.* If stopped in water, drivers should back up until well clear, using the propeller and all wheels at full throttle, and then try again at a better spot. If a stop is made practically clear of a bad surf, it probably will be necessary to winch ahead. When the DUKW is stuck in rough water, it is important to rig tarpaulins immediately and keep the pumps going.

e. When the beach is steep and the vehicle cannot make the grade, DUKW's should back out and make a new approach, shifting to first gear without stopping and before reaching the difficult pull.

f. The four hull drain valves on vehicles so equipped should be opened as soon as possible after leaving water. As soon as all water is drained out, the valves are closed. In below freezing weather, drivers must also drain propeller shaft housings immediately after leaving the water.

44. OPERATIONS ON CORAL. a. General. There are three basic types of coral formations. (See fig. 22.)

(1) *Barrier reef.* The barrier reef is usually 1 or 2 miles offshore. It may run for many miles, without a break,

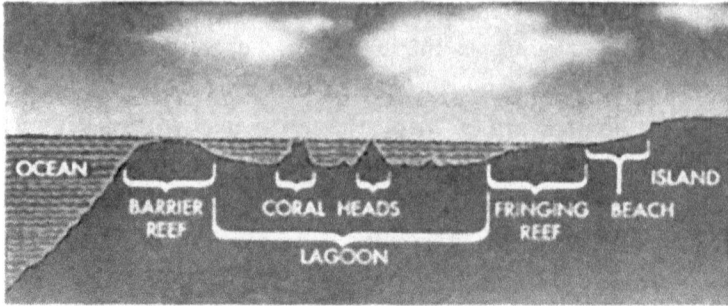

Figure 22. Typical coral formation.

at approximately the same distance from the shore, and is generally about 300 feet in width. The barrier reef acts as a breakwater and forms, between the reef and the shore, a smooth lagoon which is deep in most places. Seen from offshore, the barrier reef with its heavy surf is formidable in appearance. Actually it is less dangerous than other coral formations, since the action of the surf keeps the surface of coral formations comparatively smooth. Before crossing a barrier reef, drivers should select a point which is free from large cracks (called fissures), boulders, and loose rock; use 30 pound tire pressure; and go very slowly.

(2) *Coral head.* The coral head is usually an isolated, circular, drumlike patch up to 100 feet in diameter, frequently appearing inside lagoons. Since the water surrounding coral heads usually is smooth, their presence will not be betrayed by surf. Coral heads may be detected by the change in color of the water from a deep blue to light brown. Coral heads should be avoided whenever possible. Their surface is made up of rounded heads upon which the bottom of the DUKW may easily become stuck.

(3) *Fringing reef.* The fringing reef extends out from the shore in the form of a flat ledge a few inches below the

56

surface of the water. Its surface is composed of many sharp coral points, which can seriously damage tires unless they are inflated to 30 pounds air pressure and a very slow speed is maintained.

b. Tire pressure. For all coral operations with normal loads, tires should be inflated to 30 pounds. Tire pressure will be increased 1 pound for each 1,000 pounds of over-load. The importance of proper tire pressure is shown in figure 23.

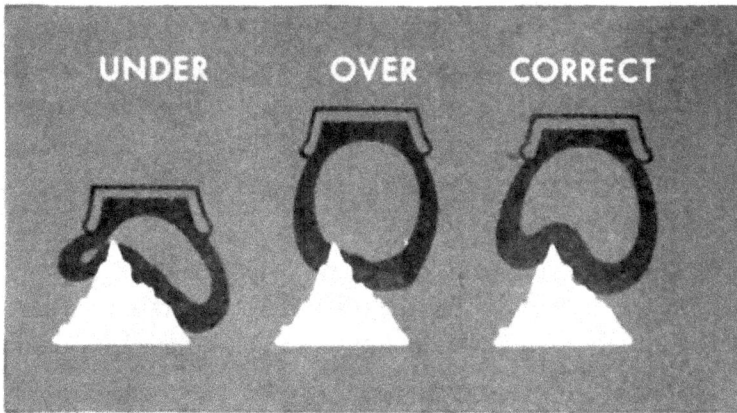

Figure 23. Thirty pounds tire pressure required on coral and sharp rock.

c. Good lookout. It is important when approaching coral infested water to maintain a good lookout. During the day time the assistant driver should go forward and stand on the bow deck. He can steady himself with a short line made fast to the davit eye. In this manner he can see the coral much better than the driver, and can direct him to the right or left, to slow down, etc., by hand signals. The water in coral areas is very clear and operators soon become experts in judging the depth of water and the character of coral reef.

d. Picking the landing point. It is important to select

the most desirable landing point, avoiding large coral heads, fissures on the edges of fringing reefs, and holes in the surface of the reef. No attempts should be made to land on a reef with an abrupt edge, since the front wheels of the DUKW will not take hold. When crossing a lagoon, the driver should avoid isolated coral head clusters by passing to either side.

e. **Approaching coral.** Upon approaching a beach where coral or rock is suspected, drivers should set proper tire pressure, place transmission in low range, first gear, and drive at the slowest possible speed. Coral can cut tires like a knife if these instructions are not followed. By approaching slowly, the operator is also less apt to damage the hull and chassis should he strike a coral head, and he is not as likely to get his DUKW jammed between two coral heads or in a fissure.

f. **Propeller.** The driver should disengage the propeller as soon as possible after the wheels have taken hold and not wait until the DUKW has reached the beach.

g. **Night driving.** Driving in darkness requires great caution. If possible, landing spots should be selected during the day, marked by range marks during the day and by range lights at night. The ranges should be carefully placed so that when the operator is running the range, that is, keeping the two ranges in line, they will lead him to the most desirable landing point on the beach.

h. **Inspection for damages.** After every landing over coral, inspection should be made of the DUKW for damages. Coral may cut tires and do other serious damage without the crew having become aware of it. Particular attention should be given the tires, front springs, brake hose, tie rods, differential housings, breather tubes, shaft housing seals, interhousing drain, propeller and propeller shaft, and strut bearings.

45. MARKING LANDING POINT. A separate landing point should be allocated to the DUKW in order to avoid the big underwater holes dug out by the propellers of the landing craft on the beaches. This landing point should be marked by some sign for the benefit of the DUKW drivers and to keep landing craft away. The sign should be illuminated at night if the tactical situation permits.

46. OPERATING FROM RAMPS OF LANDING SHIPS. For details on operating from the ramps of landing ships see paragraph 94.

SECTION VIII

TOWING IN WATER

---

47. GENERAL. DUKW's frequently are used to tow other DUKW's or small boats and landing craft. It is essential that the fundamental principles of towing be understood by all DUKW operators.

48. TOWING FROM AHEAD. a. Simple tow. To tow another DUKW or other small craft, the simplest method is to use the winch cable of the towing DUKW. The cable should be payed out until there are five full turns left on the winch drum and then should be hooked in the towed DUKW's center bow shackle at the base of the bow fairlead.

    b. Bridle tow. A bridle tow should be used when a large boat is towed or when a DUKW that is afloat is used

Figure 24. Bridle tow—use longest possible rope.

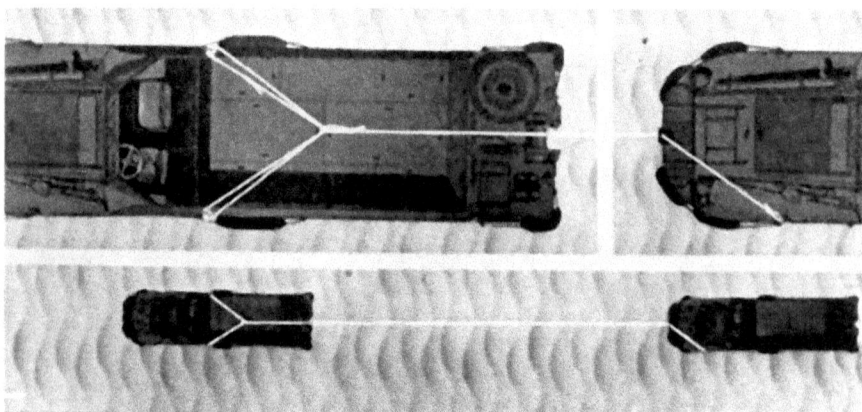

to pull a stranded boat off the beach. A bridle tow is made by attaching a rope bridle, made of 3-inch rope, to the mooring eyes. (See fig. 24.) One end of a strong tow rope, at least 60 feet long if possible, should then be tied around the loop of the bridle and the other end attached to the towed craft. This method will increase considerably the maneuverability of the towing DUKW, since the stern will not be restricted by the towing line which would limit rudder action. The DUKW crew must stand clear of the tow rope at the stern since it will sweep from side to side when turning. The rear surf board must be folded down to clear the tow rope.

49. TOWING FROM ALONGSIDE.
Towing alongside should be attempted only in smooth water. This method of towing is helpful when considerable maneuvering is necessary, when the rudder of the assisted boat is jammed, or when it is not necessary to come up on shore. Fenders should be used between the two boats. The boats should be moored together as illustrated in figure 25, showing use of breast and spring lines. The breast lines at front and rear should extend from the outside lifting eye of one DUKW to the outside lifting eye of the other DUKW (or to corresponding positions if the assisted craft is not a DUKW). Spring lines should extend from inside the front lifting eye of the towing DUKW, to the inside rear lifting eye of the assisted DUKW (or to corresponding positions if it is not a DUKW). When the craft are moving, breast lines should be slacked so there will be 1 or 2 feet between the two craft, with the bows farther apart than the sterns.

50. PUSHING FROM ASTERN.
a. The DUKW may be used to push from the stern only when the vessel to be

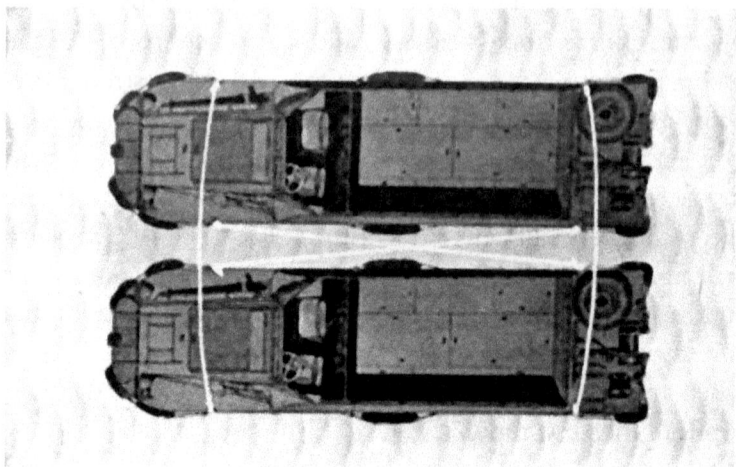

Figure 25. Towing from alongside—use in smooth water when considerable maneuvering is necessary.

pushed is large enough to present a vertical surface of sufficient height to engage the bow of the DUKW without causing damage. It is not practical to push another DUKW or any type of boat or barge which is so low that the DUKW would have a tendency to climb up on it, or to push a higher boat that has a flare or an overhang which would force the DUKW's bow down.

b. Pushing is practicable as an emergency expedient when there is not sufficient time to rig lines or when it is necessary to push a craft onto the beach, or where there is not sufficient room for towing from ahead or alongside. It is also a handy and quick method for turning a large ship in a narrow channel where one or more DUKW's can come up to the side of the bow and push it in the desired direction.

c. When possible, before undertaking any pushing operations, protection should be given the bow of the

DUKW by rigging additional fenders or mats. The front surf plate must always be folded down to protect it from damage.

51. CLEARING LINES. In all towing or pushing operations extreme care must be taken that lines are kept clear of the propeller and the wheels. This is particularly important when towing from astern. If it is necessary to slow down, the assistant driver should take up any slack in the tow rope so that it will not foul the propeller. While towing astern, the DUKW should never be put in reverse gear.

52. BEING TOWED. When a DUKW is towed, the hook in the tow cable should be hooked to the shackle below its bow fairlead. If a line is used, it should be tied through this same shackle with a bowline knot. If the line is too large to be tied through the shackle, the line should be led through the bow fairlead and tied to the left forward lifting eye. A tow rope or mooring line should never be secured to the bow fairlead or to the davit eyes.

SECTION IX

ANCHORING

53. GENERAL. There are normally two occasions during water operations when it is advisable to anchor. One is when the engine will not run and there is a possibility of the DUKW drifting into dangerous surf or away from the desired destination. The other is when the engine is able to run but the position or desired course is unknown and the driver must wait for better visibility (daybreak, lifting of fog, end of blackout, etc.) before proceeding. Anchoring by the stern is not advisable. The anchor must always be *shackled* to the cable; the hook should *not* be used.

Figure 26. Always anchor by the bow. Remember surf plate must be set up first and cable must always be shackled to the anchor.

54. ENGINE NOT WORKING. The procedure for anchoring if the engine is *not* working and the water is not over 50 feet deep is as follows:

a. First the bow surf plate is set up.

b. If it is necessary to work on the engine, the winch cable is led forward through the rear coaming fairlead, to the left of driver's compartment, through a snatch block hooked in the left forward lifting eye, and then through the bow fairlead. (See broken line in fig. 26.) The cable is brought back along the right deck to the rear and *shackled* to the anchor which is then dropped overboard.

c. If it is not necessary to work on the engine, the end of the winch cable is brought directly forward and led through the rear coaming, windshield base, and bow fairleads, after which the procedure is as in b above. (See solid line in fig. 26.)

d. Before the anchor is dropped overboard, the cable should be payed out by releasing the winch jaw clutch until only five full turns of the cable are left on the winch drum. The jaw clutch should then be engaged and the anchor dropped overboard.

55. ENGINE RUNNING. The procedure for anchoring when the engine is running and the water is not over 50 feet deep is much the same as in paragraph 54a and c above, except that first the jaw clutch on the winch is engaged, and then the anchor thrown overboard clear of everything. The cable is payed out, using the engine, until five full turns are left on the winch drum.

56. DEEP WATER. When the DUKW is anchored in water over 50 feet, the procedures in either paragraph 54 or 55, depending on the circumstances, are followed.

65

The auxiliary cable should be shackled between the anchor and the winch cable. The hook should not be used. Results are not always successful if the water is over 125 feet deep.

57. RAISING ANCHOR. The procedure for raising the anchor is as follows:

a. If the engine is not running, the crew should pull the cable in by hand, while standing in the cargo space, if possible.

b. If the engine is running, the anchor should be wound in by power. If the auxiliary cable has been used, the winch will be stopped before the end of the auxiliary cable reaches the bow fairlead. The remainder of the cable will have to be pulled in by hand.

c. If the anchor is stuck and it is impossible to raise it, a life preserver or other suitable floating object should be fastened to the winch cable outside the bow fairlead and the cable dropped by loosening the U-bolts on the winch drum. Directional bearings should be noted carefully so that the float can be located and the cable and anchor recovered when time and equipment are available.

SECTION X

MOORING AT SHIPSIDE

---

58. GENERAL. DUKW drivers must often moor at shipside while engaged in discharging cargo. In view of the relatively small cargo space in DUKW's, accurate positioning at the time of mooring is important. Mooring lines must be correctly installed at the start of an operation, and all lines must be properly adjusted, carefully tended, and maintained throughout operations. (See fig. 27.) This will assure quick, safe, and easy handling of cargo.

59. APPROACHING SHIP. When a DUKW approaches a ship to receive or deliver a load, it should take up a position approximately 100 feet off the ship's stern and on the same side where the driver intends to moor. Immediately upon receiving a signal to moor, the DUKW will

Figure 27. Mooring rig, showing spring lines and messenger lines.

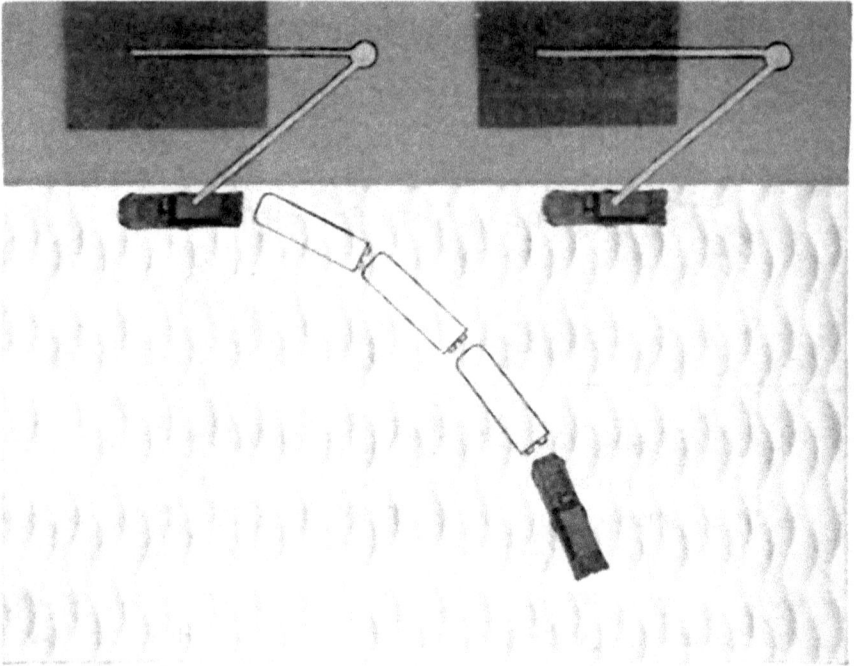

Figure 28. Proper approach to mooring position from astern.

approach the mooring position at approximately a 45°
angle from astern. (See fig. 28.) As the DUKW approaches the mooring hook, it will be slowed down and
turned parallel to the ship about 1 to 3 feet away from the
ship's side. The assistant driver, from a position between
the cab and cargo compartment, will reach the mooring
hook and quickly hook it into the DUKW's mooring eye.
(See fig. 29.) The DUKW will then be driven ahead
against the spring line with the rudder sufficiently turned
away from the ship to force the DUKW in against the
ship's side and hold it there during the loading. In mooring, it is important to approach with sufficient speed to
retain control, but slowly enough to engage the mooring
hook in the mooring eye of the DUKW before the mooring
position is passed.

Figure 29. Spring line leading astern with hook in the mooring eye. Note slack messenger line leading to ship's deck.

60. TYPES OF MOORING RIGS. a. Spring-line mooring. The DUKW is moored at shipside by a single spring line, $3\frac{1}{2}$ to $4\frac{1}{2}$ inches in circumference and about 100 feet long. The spring line is secured on the deck of the ship. (See fig. 27.) The forward end of the line carries a mooring hook which is hooked in the mooring eye of the DUKW. A messenger line, $2\frac{1}{2}$ to $3\frac{1}{2}$ inches in circumference, is fastened to the spring line 4 feet back of the mooring hook and leads to the deck of the ship directly above the mooring position. If necessary, the spring line should be parceled to prevent chafing at the ship's deck.

b. Guest-warp mooring. Where the ship is rigged with a guest warp, the individual spring lines should be attached to the guest warp rather than led to the deck of the ship. The individual spring lines can be shortened to about 50 feet in length. The messenger lines will be rigged leading to the ship deck.

c. Mooring pendant. As an expedient, if there is a fast current pushing the DUKW against the ship, it may be helpful to hook the spring line to an 8-foot wire or heavy rope pendant attached to the bow lifting eye. This pendant should be fastened to the mooring eye with a messenger line not more than 2 feet long to make handling easier.

61. RIGGING CARGO BOOMS. In loading operations, the ship's cargo booms must be rigged so that the draft centers only 5 feet outboard from ship's side when hanging from the outboard boom. (See fig. 28.) Booms over the No. 1 hatch should be rigged as far aft from the bow flare as possible. Booms over the No. 5 hatch should be rigged as far forward from the stern flare as possible. All booms should be rigged to keep DUKW's clear of dangerous hull obstructions, and to provide all possible space between adjacent berths. Double rigged hatches should be worked from opposite sides of the vessel except when rough water necessitates operation from the lee side only.

62. MAKING A LEE. In extremely rough weather, the ship will do all possible to provide a lee of approximately 20°. (See fig. 30.) DUKW's will moor on the lee or sheltered side when practicable.

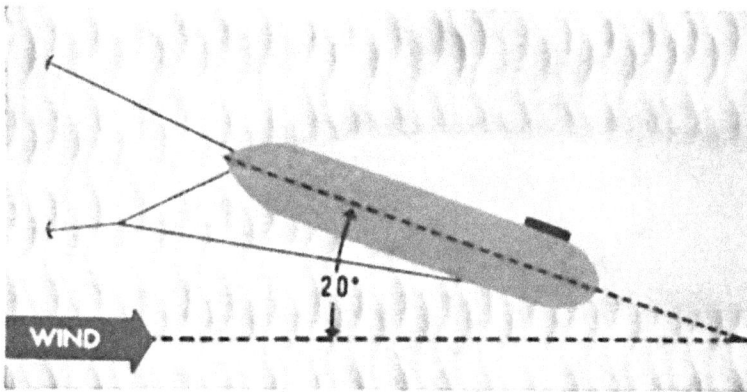

Figure 30. Ship making a 20° lee.

63. ADJUSTING SPRING LINES. When the first draft of cargo is put over the ship's side, the spring lines should be adjusted so that loads will hang over the center of the DUKW cargo space when the DUKW being driven forward at one-third throttle (1,000 rpm). (See fig. 31.) When this adjustment has been made, it should remain satisfactory for long periods of the loading operation.

Figure 31. Spring lines are adjusted so the load comes down directly over the center of the cargo space.

a. When a draft is received of such weight that is difficult for the DUKW crew to place it in the desired location in the cargo space, the DUKW's position can be adjusted by changing the engine speed. For example, if the cargo hangs at the center of the cargo space when the engine is running at 1,000 rpm, idling the engine will normally drop the DUKW back 3 or 4 feet, and the cargo will come directly to the forward end of the cargo space. In order to place a heavy draft at the rear end of the cargo space, the engine speed should be increased to 1,500 rpm, pulling the DUKW forward about 3 feet and bringing the load directly to the desired space.

72

b. As a rule, in smooth water with little or no current, low engine speeds will hold the DUKW's in position. In rough water or with a strong current, increased engine speed will be necessary to keep the spring line reasonably taut and the DUKW alongside the vessel.

64. FENDER PROTECTION. If a ship has a guardrail or rough guards over hull outlets, rope or mat fenders should be rigged by the ship's crew to protect the DUKW's. A floating log fender is very practical. If a ship has reasonably smooth sides, the DUKW's fenders should give sufficient protection. See paragraph 127 for instructions in making a fender.

65. POSITIONING UNDER FLARE. a. When there is considerable flare abreast the No. 1 and No. 5 hatches, it will be necessary to swing the DUKW's stern *out* to get the cargo space under the load. (See fig. 32.)

Figure 32. Where there is considerable flare abreast the No. 1 and No. 5 hatches, it will be necessary to swing the stern out to receive the load.

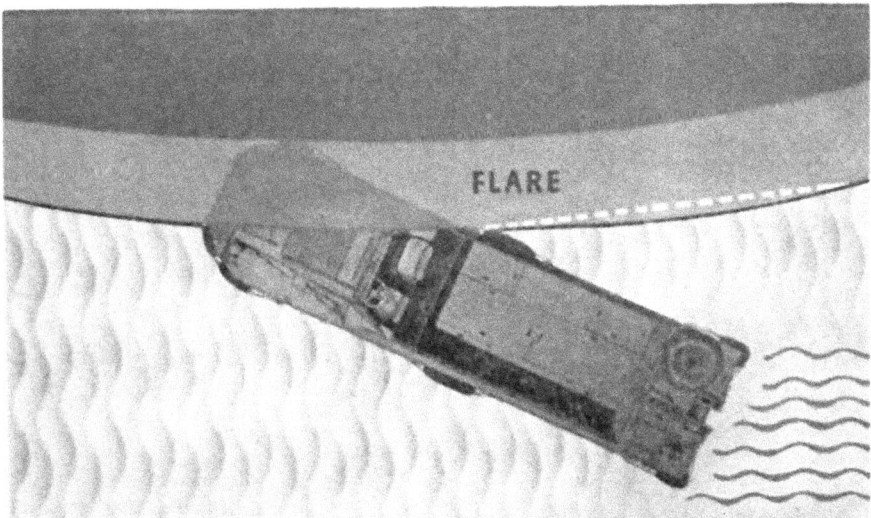

FLARE

In cases of extreme flare at No. 5 hatch, particularly in rough water, the spring line should be hooked to the inside bow lifting eye, using little or no power after picking up the hook. Under these conditions, an additional stern line may be helpful but will need constant attention because of its unfavorable angle.

b. Another method for positioning under a flare, if necessary, is for one DUKW to moor in the usual manner and have the DUKW that is to receive the cargo tie up along the outside of the first DUKW.

66. MOORING HOOKS. It is not practical to moor without hooks or with inadequate hooks which may give way at a critical time. A suitable mooring hook has been furnished with each DUKW above serial number 353–2006. Some hooks are fitted with a tongue or latch which makes them, in effect, large snap hooks. This latch should be removed, leaving the hook open. Where these hooks are not available, any one of several alternate types, as illustrated in figure 33, may be used. To facilitate night operations, hooks should be painted with phosphorescent or white paint.

Figure 33. Alternate types of mooring hooks that may be used.

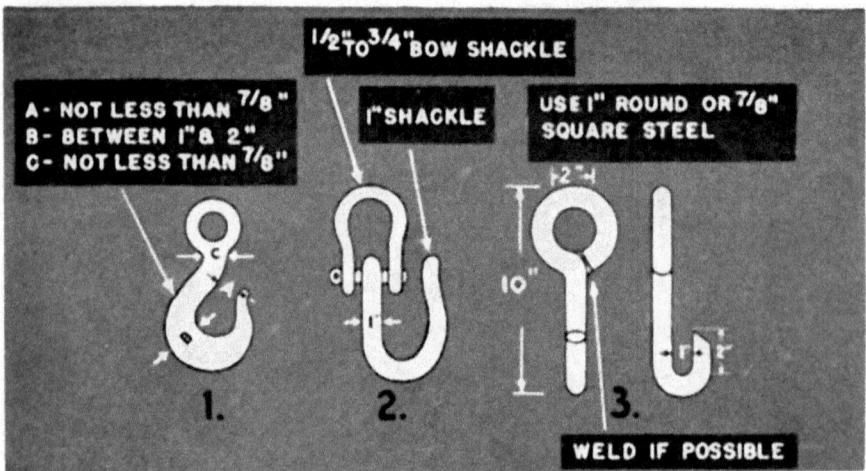

67. STERN-TO-BOW CURRENTS. When the current or tide runs by the ship from stern to bow because the ship is aground or moored to a dock, the mooring rig will be reversed and spring lines will be rigged to lead *aft. DUWK's will always come in bow heading into the current.*

68. SURF PLATES AND SHIELDS. Whenever the DUKW is going to moor, the bow surf plate will be lowered since, when it is set up, it extends beyond the bow and bow fenders and will be damaged. (See fig. 34.)

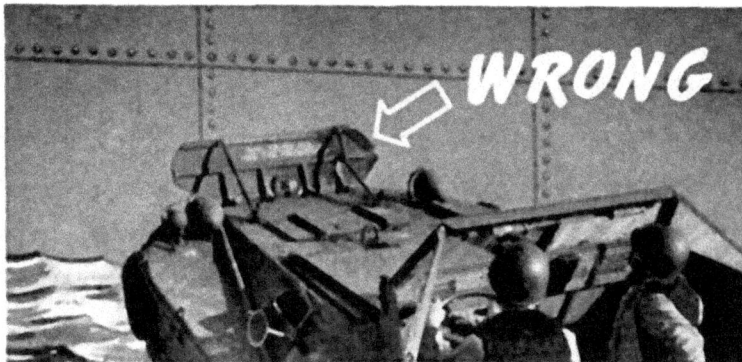

Figure 34. Failure to lower surf plate before mooring may cause damage.

The plywood rear surf shield, on DUKW's so equipped, should also be folded down before mooring to prevent its being damaged.

69. WHEN ENGINE STOPS. If the engine stops and cannot be started while the DUKW is moored, the driver should cast off immediately, even if this necessitates cutting the spring line with a knife or an axe. After the

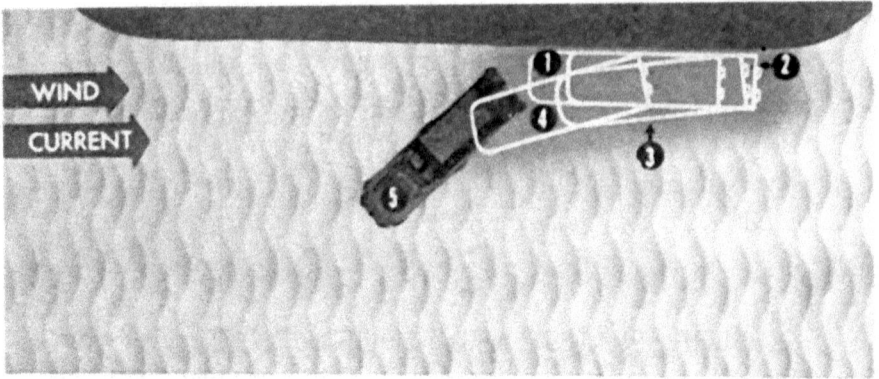

1. Moored alongside.
2. Drop back sufficiently to permit unhooking spring line.
3. Push off by hand before starting ahead.
4. Drive ahead, turning away gradually so stern does **not** swing against ship, as this would tend to pull the DUKW back alongside again.
5. All clear—take desired course.

Figure 35.   Leaving a mooring under favorable conditions.

1. Moored alongside.
2. Drive ahead, with steering wheel turned **toward** ship, forcing stern out.
3. When this position is reached, unhook mooring line and back away.
4. When this position is reached, turn **away** from ship and drive ahead.
5. All clear—take desired course.

Figure 36.   Leaving a mooring under unfavorable conditions.

DUKW drifts clear of the ship, the signal for assistance should be displayed. (See paragraph 98.) The DUKW should await a tow and, if no tow is in sight, should anchor.

70. PROCEDURE FOR LEAVING A MOORING. a. Favorable conditions. Under favorable conditions, the clutch should be disengaged and the spring line unhooked as soon as it is slack. The bow should be pushed off by hand and, when the bow has swung away from the ship, the clutch should be engaged to drive the DUKW ahead. The driver must not swing away sharply or the stern will hit the ship and pull the DUKW back alongside. (See fig. 35.)

b. Unfavorable conditions. If the tide or other conditions make it impossible to leave the ship's side in the manner described above, the steering wheel should be turned toward the ship before the DUKW is unhooked, and the propeller should be driven ahead to swing the stern out. The DUKW should then be unhooked and backed away with the propeller in reverse and the wheels in high transfer. When well clear, the DUKW should be turned away and driven ahead. (See fig. 36.)

SECTION XI

PILOTING

71. GENERAL. A working knowledge of the fundamentals of piloting is essential to the efficient operation of DUKW's on the water. Piloting is the technique of using visual observation, ranges, the compass, charts, and soundings in order to get from place to place while afloat. When operating in quiet water where visibility is good, the piloting of DUKW's is simple—the driver sees his objective and steers for it. However, such ideal conditions are not always present.

72. CURRENTS AND TIDES. a. In almost all water operations, current caused by tidal movement or by the proximity of rivers will be encountered. The speed and effect of the current must be allowed for when mooring, landing, or when holding a course between two points. Figure 37 shows the effect of a side current and the correct way to allow for it in traveling between two points. Proper allowance for current becomes vitally important when visibility is poor, and also when a definite landing position must be accurately approached.

    b. The depth of water along the seacoast in almost all parts of the world varies according to a predictable 12-hour cycle. This is called tide. There normally will be a high tide twice each day. The difference between the depth of water at high tide and at low tide is known as the range of the tide. Average range is about 5 feet, though it varies from less than a foot in some places to over 30 feet in others. It is important to know that certain conditions which may exist for one trip may be different later because of a change in tide. Shoals which may safely be crossed at high tide may be dangerous at low tide. A shore line may easily be crossed at one stage of the tide and be quite difficult at another.

73. UNDERWATER OBSTRUCTIONS. The most difficult condition under which DUKW's can operate is that of driving on land covered by water. In deep water, normal boat operating technique can be applied. On dry land the problem is one of skillful truck driving. But when wheels or bottom touch ground covered by water, the operation of the DUKW becomes complex. It is important, therefore, for drivers to operate in deep water until the landing point is reached. Intervening shallows should be crossed as directly as possible. The DUKW should be driven clear of the water and onto dry land as quickly as possible. This means that the location and depth of water over shoals, reefs, and bars must be determined. Since these obstructions frequently cannot be seen because of darkness or muddy water, they must be located by referring to fixed points on shore, to buoys, or to other markers in the water.

74. SOUNDING. Where charts are not available and where the tactical situation permits, reconnaissance of *underwater* conditions should be made. In the neighborhood of a proposed landing site, and where reefs or shoals are close by proposed operations, this reconnaissance is particularly important. The depth of water, up to 8 feet, can be measured by using the boat hook. When making these soundings it is possible by the "feel" of the boat hook to determine the consistency of the bottom. Soft mud or extremely soft sand should be avoided.

75. PILOTING BY RANGES. The easiest way to sail the shortest course between two points is by the use of ranges. Ranges are usually artificial markers placed on the shore in pairs, one some distance behind and higher than the other. They may, however, be arranged in different manners, as indicated below. They may also be such natural items as trees, rocks, hills, or anchored vessels. Different arrangement of ranges are (fig. 37) —

    a. Two markers ahead. One marker should be rather near the shore line and the other farther away and preferably higher.

    b. One marker ahead, the other astern. For example, the ship could be used as the marker ahead and the place where the DUKW went into the water as the marker astern.

    c. Two markers astern. This is the same principle as the two markers ahead.

As long as the two markers chosen are kept in a direct line, the DUKW will be holding the course. Regardless of the method used, bearings on the markers should be taken often to note the effect of the current, and necessary allowances should be made. When the current carries the DUKW off its course, the driver should steer into the current at a sufficient angle to keep the two range markers in line.

76. MAGNETIC COMPASS. a. General. The magnetic compass is used to orient the various points on a chart or map. The compass installed on the DUKW will be useful only if kept in adjustment. The great amount of steel and the number of electrical parts close

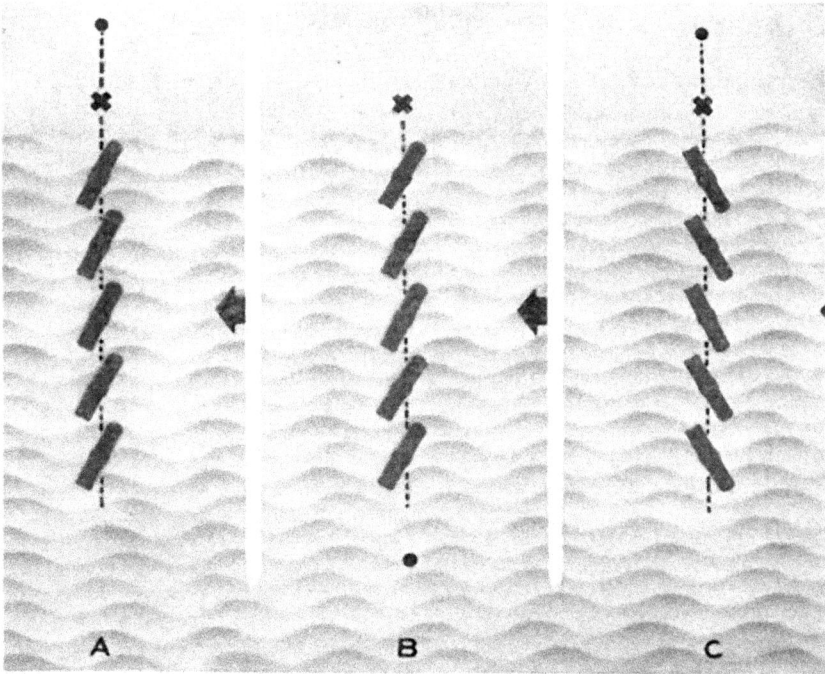

*Figure 37. Piloting by ranges: (a) Two range marks ahead;
(b) One range mark ahead and one astern; (c) Two range
marks astern. (Note that the DUKW is headed up stream to
compensate for the current.)*

to the compass make frequent checking and corrections
necessary.

b. Adjustment. (1) To adjust the compass, the driver
should first mark out accurate north-south and east-west
lines on a level portion of the ground. This can be done
by driving three stakes into the ground, a center stake,
and a north and a west stake each about 40 feet away
from the center stake. (See fig. 38.) The position of the

north and west stakes should be determined by using a pocket compass, remembering to keep the compass at least 10 feet away from any metal object. The compass from a DUKW should not be used since it has built-in magnets and may be in error when removed from the vehicle.

(2) After the direction lines are marked out accurately, the DUKW should be driven so that its left side is along the north-south line, heading directly north. The N-S (north-south) corrector on the compass should then be adjusted until the compass reads north.

Figure 38.   Procedure for lining on directional stakes when adjusting the compass.

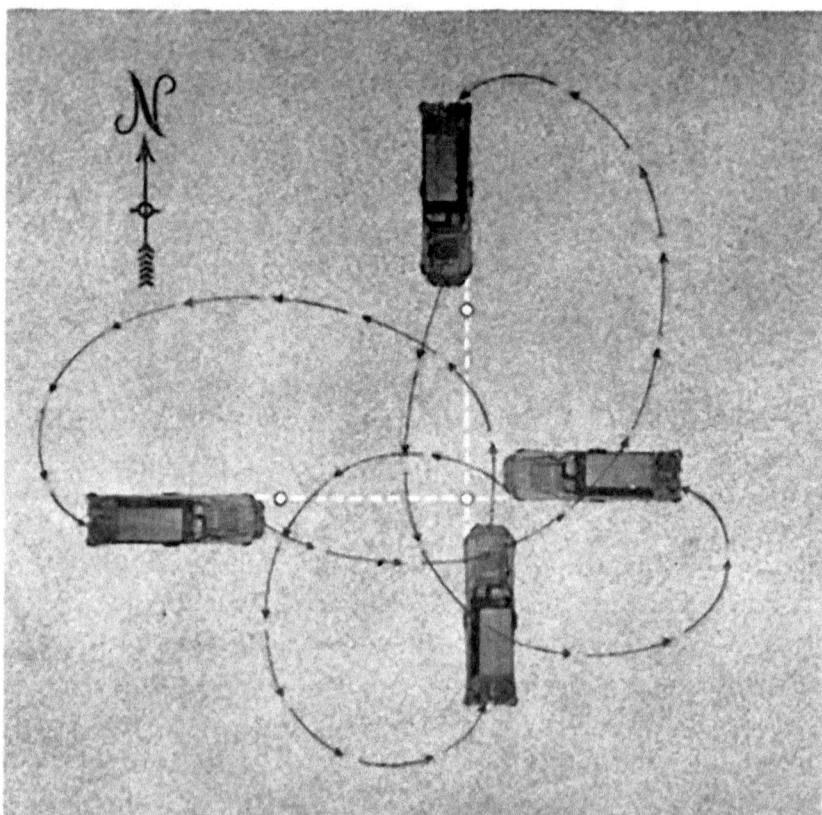

(3) The DUKW is next headed east, and the E-W (east-west) corrector is adjusted until the compass points east.

(4) The DUKW is then driven so that it is pointing south along the north-south line. If the compass does not then point due south, the driver will proceed as follows:

(a) He will note the number of degrees of error and divide by two.

(b) He will then adjust the compass toward the south by the amount determined in (a) above.

(5) Next, driving so that the DUKW faces west, the driver follows the procedure in (4) above except that the adjustment will be made toward the west.

(6) The DUKW is then driven so that it points in any one of the cardinal (north-south-east-west) directions. The driver will then proceed as in (4) above, with adjustments being made as the DUKW heads in all four cardinal directions in succession. This procedure should be continued until the errors remaining in the compass are as nearly as possible the same for the north as for the south readings and the same for the east as for the west.

(7) Any errors which remain in the compass should then be recorded on a deviation card. Errors are determined by noting the difference between the compass reading and the true reading. If the compass reading is smaller than the true reading, the difference is plus; if the compass reading is larger, the difference is minus. For example, if the north compass reading is 354° (true north is 360°), the error is plus 6°. Also, if the south compass reading is 175° (true south is 180°), the error is plus 5°. A sample deviation card might be as follows:

Heading	N	E	S	W
Deviation	$+6°$	$-2°$	$+5°$	$-2°$

(8) Deviation for intermediate points of the compass can be estimated from the known deviation of the north, south, east, and west readings by interpolation.

(9) If the compass is equipped with additional adjustments, instructions issued by the manufacturer should be followed.

77. LOG. The log is the name applied to the written record of a trip, particularly one on which piloting is involved. Where the trip is lengthy or of a complex nature, an accurate written log should be kept, noting the time of leaving ship and shore, the passing of any special markers, etc. Compass courses should be noted even if it is not necessary to check the bearings at the time. This is important because it may be necessary to repeat the trip later after either a fog, rain, darkness, or a blackout has obscured guiding markers. The driver can then refer to the written log for his course, the running time, etc.

78. ORIENTATION. For various methods of orienting himself, the amphibian truck driver should refer to FM 21–25 and TB 21–25–1, "Elementary Map and Aerial Photograph Reading."

79. CHARTS. Charts are the marine equivalent of maps. Where under water shoals, reefs, and bars must be avoided, their location can be learned from studying a chart of the locality involved. Explanations in the margin of the chart provide keys to various markings. Distances are usually shown in nautical miles. Current information is based on knots which are nautical miles per hour.

80. KNOTS AND NAUTICAL MILES. A nautical mile is a unit of measure of distance used on marine charts. A knot is a unit of speed equivalent to one nautical mile per hour. Knots and nautical miles can be converted to land miles by multiplying by $1\frac{1}{7}$. Land miles can be converted into nautical miles by multiplying by $\frac{7}{8}$. DUKW speed and endurance figures are based on land miles.

81. PLOTTING COURSE. Plotting the course is the procedure used to determine the compass heading which the driver must steer when it is necessary to make corrections for variation, deviation, and, possibly, current. The procedure for making these corrections consists of performing, on a chart of the area under consideration, the four steps outlined below. (See fig. 39.)

a. Determining true heading. True heading is the direction, in degrees, that the DUKW is to be steered in order to get from an embarkation point to a landing point. To obtain this true heading, the first step is to draw a straight line between the embarkation point and the landing point. Next, a line parallel to this line is drawn through the center of the compass rose appearing on all charts. The true heading will be that degree on the outer rim of the compass rose lying in the direction of the landing point.

b. Finding magnetic heading. Magnetic heading is determined by adding variation to true heading where variation is westerly. Where variation is easterly, it is subtracted. A memory aid phrase for this rule is "West-Best-Add, East-Least-Subtract." Variation is the difference between magnetic north and true north expressed in degrees. It differs depending upon geographic location and is indicated on all charts.

c. Finding compass heading to steer. Compass heading to steer is determined by applying deviation to the magnetic heading. Deviation is the effect on the compass caused by magnetism within the DUKW. Paragraph 76b describes the method for eliminating all possible deviation and recording remaining errors on a deviation card. The deviation is added or subtracted depending upon whether the error is plus or minus. The results of this adjustment will give the compass heading to steer.

d. Allowing for current. Due allowance must be made for current in a body of water to be crossed unless the current is running in the same direction as, or in the opposite direction to, the DUKW's course. It is necessary to make this allowance as the first step in the procedure for plotting the course. This step should then be followed by the other three steps. In order to allow for a current, a line is drawn from the embarkation point in the same direction as, and parallel to, the average current. The length of this current line should equal the distince the current will run in 1 hour. The next step is to draw a line from the end of the current line to the line drawn between the embarkation point and the landing point. The length of this last line should equal the distance the DUKW will run, in still water, in 1 hour. Its direction should be used for determining the true heading. True heading must then be corrected for variation and deviation to find compass heading to steer as outlined in b and c above. (See fig. 39.)

82. EXAMPLE OF PLOTTING COURSE. a. It will be assumed that, in the example shown in figure 39, there is a current of 1.8 knots for which due allowance must

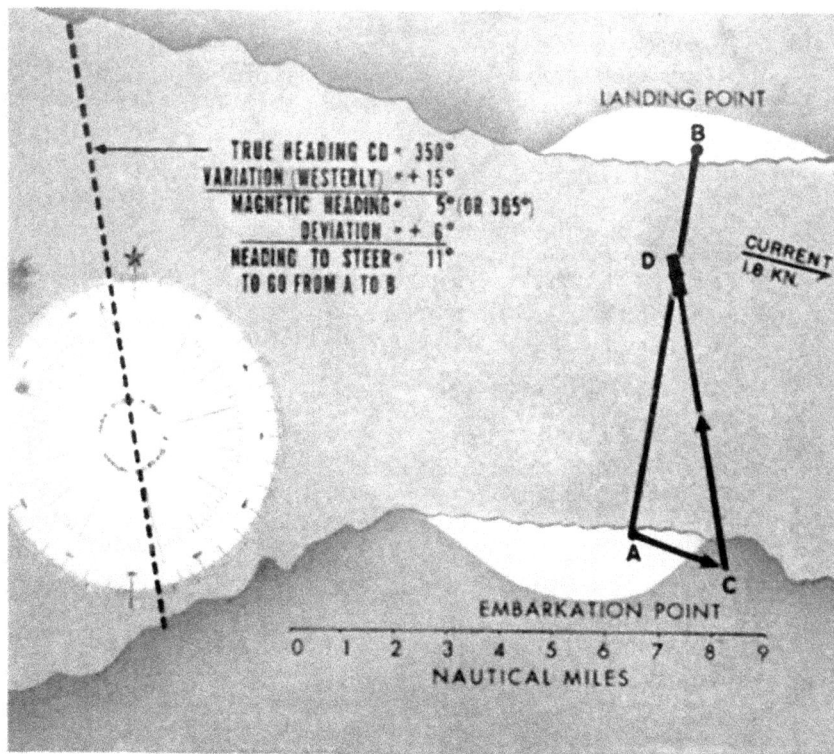

TRUE HEADING CD = 350°
VARIATION (WESTERLY) = + 15°
MAGNETIC HEADING = 5° (OR 365°)
DEVIATION = + 6°
HEADING TO STEER = 11°
TO GO FROM A TO B

LANDING POINT

B

CURRENT
1.8 KN.

D

A

C

EMBARKATION POINT

0 1 2 3 4 5 6 7 8 9
NAUTICAL MILES

Figure 39. How to determine the compass heading to steer, allowing for current, variation, and deviation.

be made. First, a line is drawn from embarkation point *A* to landing point *B*. Next, line *AC* is drawn from the embarkation point in the same direction as, and parallel to, the current. The length of line *AC* is equal to the distance the current will run in 1 hour, and is determined by measuring off 1.8 miles on the chart's scale. A line equal in length to the distance the DUKW will run in 1 hour, in still water, is drawn from point *C* to line *AB*. By drawing a line parallel to line *CD* through the compass rose, the true heading of 350° is found.

b. The compass rose on the chart shows that there is a westerly variation of 15°. In order to find the magnetic

87

heading this is added to the true heading ($350° + 15°$ = $5°$ or $365°$).

c. The deviation card shows an error of plus $6°$ on a north heading. This is added to the magnetic heading to get compass heading to steer ($5° + 6° = 11°$). Thus, steering the DUKW on the compass heading of $11°$ will take it direct from embarkation point *A* to landing point *B*.

Figure 40. Method of determining privileged and burdened vessels by use of left hand. Vessels in the space between the thumb and index finger are privileged vessels and have the right of way.

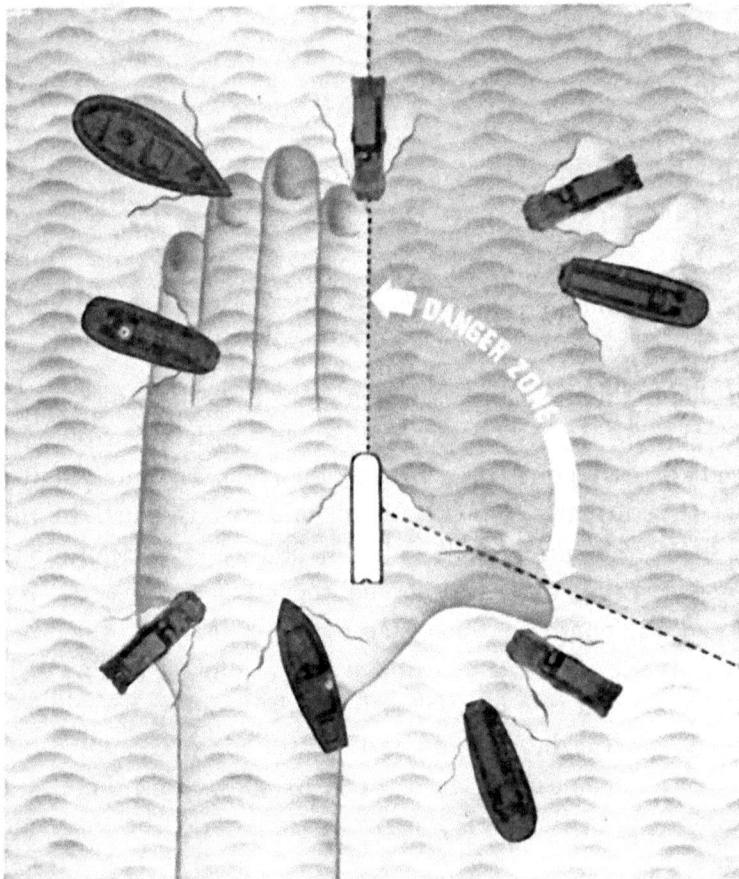

83. RULES OF ROAD AT SEA. Rules of the road are established to prevent collision at sea. They apply to all vessels, including DUKW's, under way—that is, all craft not anchored, aground, or made fast to shore. Drivers of DUKW's, therefore, must be familiar with the following basic rules. See figure 40 and FM 55–130.

a. When two boats are approaching each other head on, or nearly so, it is the duty of each to swing to the right (starboard). Each boat therefore, will pass on the left (port) side of the other. One short blast of the whistle or horn is given by each. In other words, each boat keeps to the right.

b. When two boats are approaching each other at right angles or obliquely, so as to involve risk of collision, the boat which has the other on her left (port) side is the privileged vessel and holds her course. The other which has the boat on the right (starboard) side is the burdened vessel and keeps out of the way. That is, the vessel approaching from the right, has the right-of-way. The privileged vessel gives one blast, which is answered by one blast from the burdened vessel.

c. When two boats are approaching each other, and it is considered more suitable to pass right side (starboard) to right side, the privileged boat swings to left (port) giving two short blasts. This signal is answered with two short blasts by the burdened boat which also swings to left (port).

d. When a boat is intending to go astern (in reverse) she signifies her intention of doing so by giving three short blasts of her whistle or horn.

e. At night, when the tactical situation permits, boats carry a green light on the starboard side (right) and a red light on the port side (left).

f. A DUKW coming into the shore, and while still in

the surf, has the right-of-way over a DUKW that is about to enter the water or one near the same point on the beach.

g. When operating in a river, canal, channel, or harbor entrance, the DUKW keeps to the right, whenever possible.

h. When one boat is overtaking another headed in the same direction, and wishes to pass on the left side (port) of the vessel ahead, she will give two blasts to signify her intention. If she desires to pass to the right side (starboard), she will give one blast. Either signal is answered by the same signal from the vessel ahead. In any case the overtaking vessel must keep clear.

i. When operating in crowded harbors and channels, DUKW's will keep out of the way of larger vessels, since the DUKW can turn much faster and can operate in much more shallow water than a large ship.

j. When a collision appears imminent at night and the tactical situation permits, all available lights should be flashed to attract the attention of the approaching vessel and to show the DUKW's course and position. However, lights should not be held in such a way as to blind personnel on the approaching boat.

SECTION XII

CARGO HANDLING

---

84. PAYLOAD. Determining the correct weight of payload to be carried is one of the most important considerations in handling cargo by DUKW's. This weight must be governed by operating conditions and normally should not exceed 5,000 pounds. In determining payload, *all* personnel, personal gear, spare gasoline, oil, water, parts, dunnage, nets, pallets, and all items that are not standard parts of a DUKW must be counted as part of the payload. This combination of persons and items usually weighs from 500 to 1,500 pounds; 1,000 pounds should be considered average. However, in combat emergencies and when conditions are *favorable* or *ideal,* as indicated below, a payload of between 5,000 and 10,000 pounds may be carried. More than 10,000 pounds, however, should never be carried under any circumstances.

    a. DIFFICULT conditions. Under *difficult* conditions, payload should normally not exceed 5,000 pounds. The term *difficult* will apply to all operations involving use of untried landing sites, unless reconnaissance has definitely indicated conditions to be favorable. It also applies when the surf at shore is over 3 feet, the wind over 15 miles per hour (white caps), or when the wave height at shipside is over 3 feet. In addition, it applies to operations involving coral, very soft sand, steep landings, mud, and steep hills, or when the land distance to the dump is over 6 miles.

b. FAVORABLE conditions. Under *favorable* conditions and combat emergency, payload may total 7,500 pounds. The term *favorable* will apply to operations involving use of reconnoitered landings where reasonably smooth and firm terrain is available, where surf at shore is less than 3 feet, wind is less than 15 miles per hour (no white caps), and waves at the shipside are less than 3 feet. The landing point on shore must also be negotiable in low-range second gear and the land distance to dump must be less than 6 miles, with only moderate hills.

c. IDEAL conditions. Under *ideal* conditions and combat emergency, payload may total 10,000 pounds. The term *ideal* will apply to daylight operations over a smooth, firm, and gradually sloping landing point, negotiable at a minimum of 30 pounds tire pressure; surf at shore must be less than 2 feet, the wind less than 10 miles per hour, and waves at shipside less than 1 foot; land route to dump must be reasonably level and smooth and less than 3 miles long; and water hauls must be less than 1 mile long.

85. TIRE PRESSURE. For payloads exceeding 5,000 pounds, tire pressures recommended on dash instruction plates, stencils, and in various manuals should be increased 1 pound for each 1,000 pounds overload. Minimum pressure for very soft sand with 5,000 pounds payload will be 12 pounds.

86. LOAD MARKINGS. a. To help control loading, there are three guide lines on each side of the stern of most DUKW's which indicate the water line for the three types of operating conditions. (See fig. 41.) The lowest lines

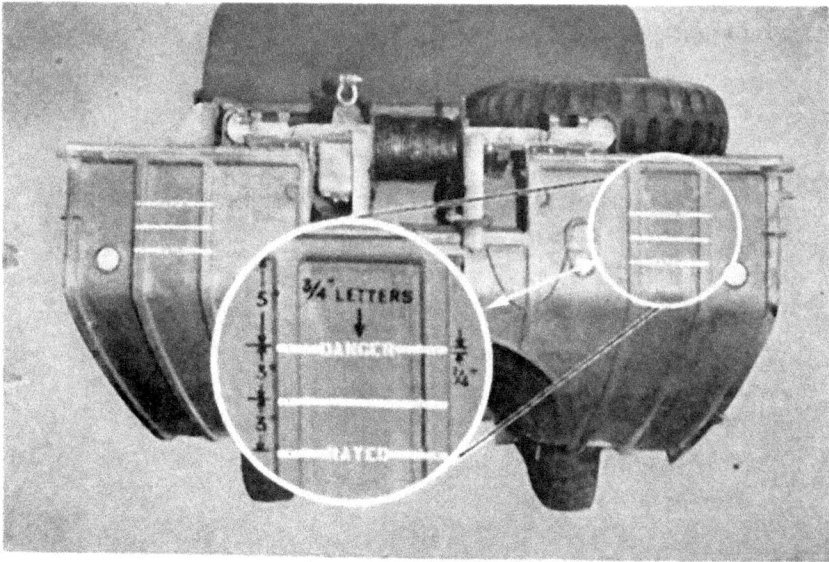

Figure 41. Load when afloat is indicated on stern load markings.

marked "Rated" are for *difficult* conditions, middle lines for *favorable* conditions, and the upper lines, marked "Danger", for *ideal* conditions. A load that puts the "Danger" lines below the water should never be carried.

b. DUKW's that do not have these load markings should be so marked with nonlustrous yellow paint strips ½ inch wide on the right and left sides of the stern. The top lines marked "Danger" should be painted 5 inches from the under side of the deck guard rail. The middle lines should be 3 inches below the top lines. The bottom lines marked "Rated" should be 3 inches below the middle lines. The lines are measured from the top of one line to the top of the next line. Lettering should be ¾ inches high.

87. TRAINING. Personnel in training should practice operating the DUKW's under conditions and payloads set forth above.

88. DUNNAGE. Dunnage must be used to protect the floor of the DUKW if the load is of a type that places considerable weight on a small portion of the cargo floor. In some cases, dunnage is used to protect the cargo, particularly when more than one layer of cargo is carried. Dunnage is also used to block around cargo that might shift during the trip over water or land. Timber, pallets, nets, steel matting from air strips, boughs, or similar available material may be used as dunnage. The use of dunnage is more important in rough water since the motion of the DUKW and the ship causes drafts to land more heavily in the cargo compartment and increases the possibility of the cargo shifting.

Figure 42. Crew guiding load to proper position in cargo space.

89. POSITIONING OF CARGO. Normally, cargo weight should center approximately 2 feet forward of the middle of the cargo space. The draft should be lowered directly over the center of the cargo space. (See fig. 42.) When cargo is of a type that completely fills the cargo space, it may not be possible to place the cargo in this position. However, it is almost always possible to pile the cargo higher at the forward end. Heavy cargoes must be placed forward of the center, otherwise the DUKW may ship water over the stern and be swamped. In addition, if too much weight is in the stern the water speed is reduced and, on land, mobility and tire life are both reduced. Lateral position of cargo should be centered so that the DUKW will remain level.

90. NORMAL LOADING PROCEDURES. a. Net loads. Generally, most cargo is handled in nets without the use of dunnage. Nets may be used for all articles that are small enough not to warrant the use of special equipment, and yet will not fall through the mesh of the nets. DUKW's will normally carry one, two, or three net loads, depending on net load weights, and on operating conditions. If one net load only is to be carried, it should be placed in the forward part of cargo space. (See fig. 43.) If two net loads are to be carried, the first draft should be placed in the rear end of the cargo space and the second in the forward end. Any overhanging cargo should be allowed to overhang in the forward end. When three net loads are to be carried, the first load should be placed in the forward end, the second in the rear, and the third on top of the first at the forward end. The third net will have to be lashed carefully so that none of its contents will be lost. Nets often have round or square pallets, called pie plates, placed in their centers. A new type of cross net with a square pallet is referred to in paragraph 128.

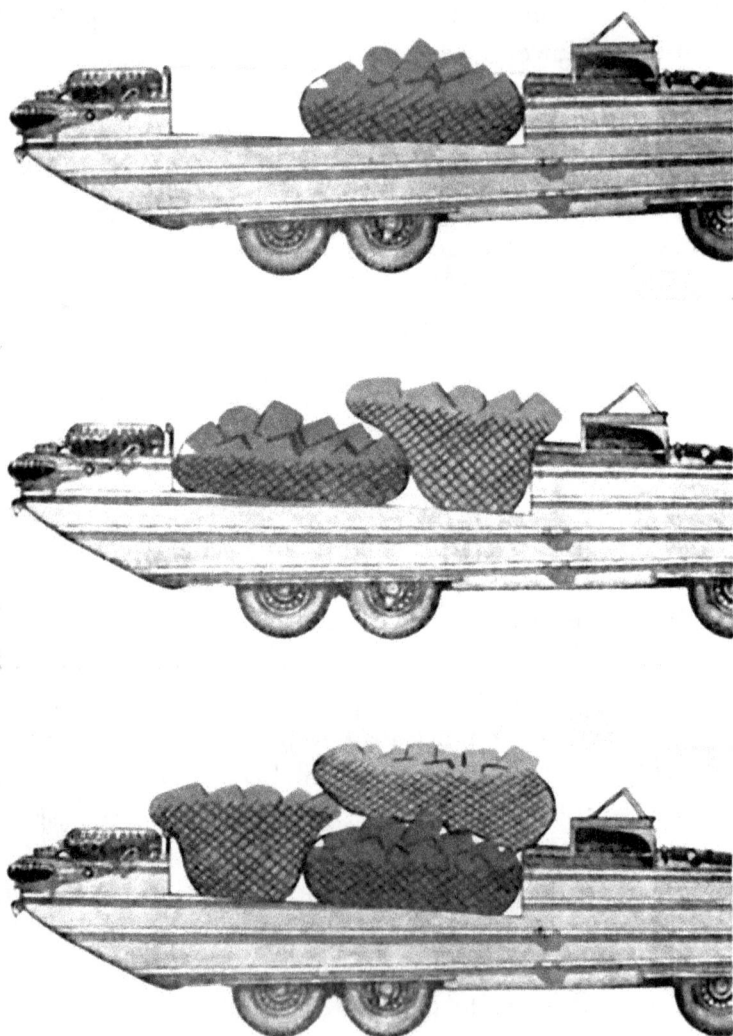

Figure 43. Positioning of one, two, or three nets in cargo space.

b. Pallet loads. Pallet loads are used almost as frequently as nets, usually without the use of dunnage.

(1) Where loads are heavy and sufficient pallets are on

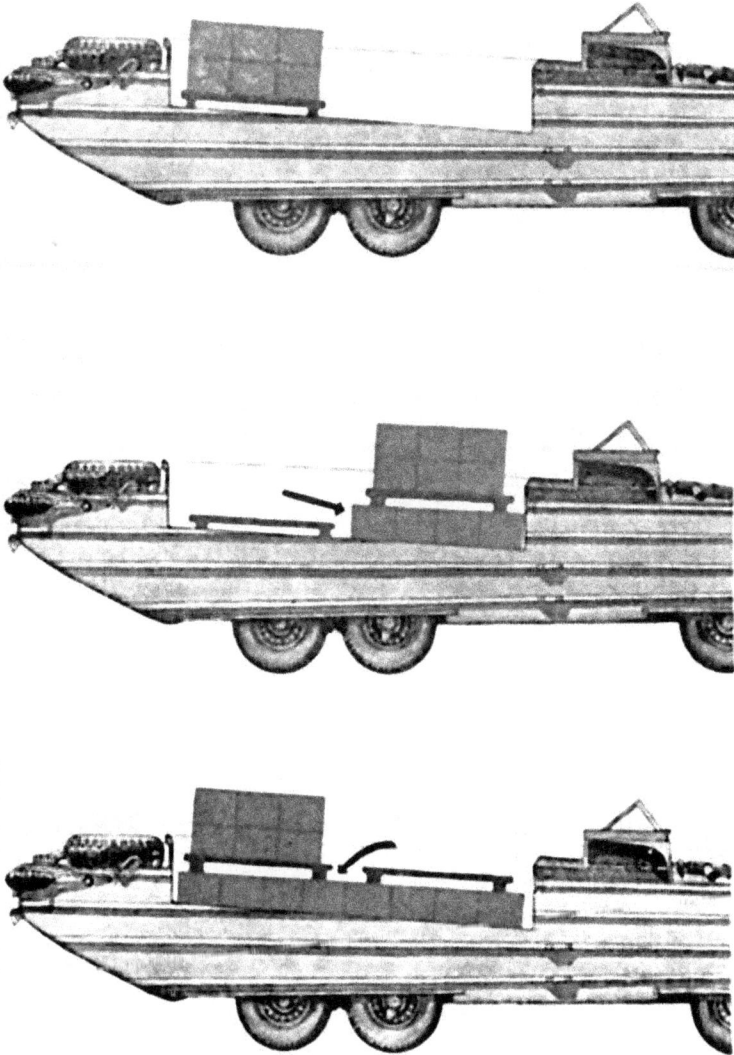

Figure 44. Procedure for unloading several pallets in cargo space.

hand, pallet loads may be carried intact. In this case, the first pallet is landed in the forward end of the cargo space, crosswise if length will permit. One or two additional pallet loads should be placed behind the first one until the cargo space is full or the desired load is aboard. When pallet loads are sufficiently secure, and conditions permit, a second layer may be carried with the loading again starting at the forward end.

(2) If pallet loads are light and there is a shortage of pallets, the first pallet should be landed at the rear end of

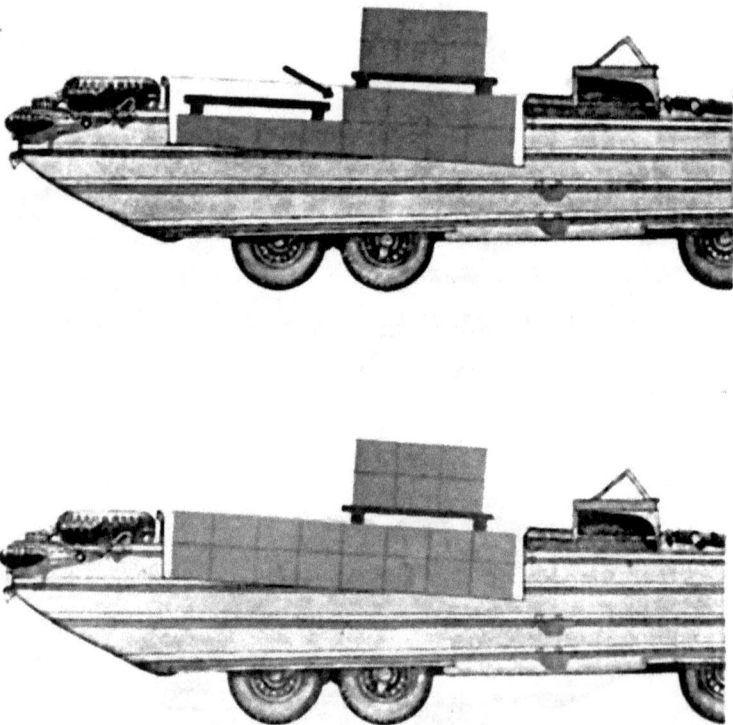

Figure 44. Procedure for unloading several pallets in cargo space.
Continued.

the cargo space. After unhooking the draft, the material should be taken off the pallet and loaded evenly in the forward end of the cargo space. The second pallet should then be landed on top of the cargo at the forward end of the cargo space. When the second pallet is unhooked, the lifting gear should be hooked into the empty first pallet, which is hoisted back aboard the ship. At this time, the second pallet will be unloaded, filling the rear part of the cargo space. The third pallet load will be landed on the cargo at the rear end and the procedure will be continued until the DUKW has been loaded. (See fig. 44.) To save time, the last pallet should not be unloaded until the DUKW reaches the dump.

c. Bound pallets. Bound pallet loads are used in some operations, usually without the use of dunnage. Each of these pallet loads usually weighs approximately 2,500 pounds regardless of the type of cargo, and should be carried intact. The first pallet should be landed at the forward end of the cargo space, crosswise if possible. Additional pallets should be loaded behind the first until the desired load is aboard. If conditions permit, a second layer may be carried, with loading again starting at the forward end of the DUKW.

d. Fuel drums. The 55-gallon fuel drums weigh approximately 400 pounds each and are handled best by using a cluster of six sets of barrel hooks. (See fig. 45.) Six or eight small pieces of dunnage are required in loading these drums. The first six drums should be set on their sides crosswise in the forward end of the cargo space. Immediately after being unhooked, the drums should be rolled against the forward cargo bulkhead, and small pieces of dunnage placed behind the rear drums to prevent them from rolling. The second cluster of six drums is placed in the rear half of the cargo space. They should be rolled

Figure 45. Drums should be handled by a cluster of six pairs of barrel hooks.

Figure 46. Positioning of 55-gallon drums under various conditions.

IDEAL ○   FAVORABLE ●   DIFFICULT CONDITIONS ●

against the forward drums and blocked. For *difficult* conditions, only one layer of drums should be loaded. (See fig. 46.)

(1) Under *favorable* conditions, an additional cluster of six drums is placed on top of the drums at the forward end of the cargo space. It is extremely important that suitable dunnage be wedged behind these drums to prevent them from rolling backward, particularly when climbing up on the beach or up a steep hill on land. If the drums should roll back, they might smash the rear coaming and be lost over the stern.

(2) Under *ideal* conditions, a fourth cluster of six drums may be placed at the rear end of the cargo space. They should be placed up against the rear coaming to prevent them from rolling.

e. Boxed ordnance parts. Dunnage is necessary in loading boxed ordnance parts only as explained below. The boxes normally will be loaded singly or in groups using rope slings. Where enough slings are available, they should stay with the boxes during the trip to the dump for use in unloading. If there are not enough slings, 2 by 4's or other dunnage of similar size must be used in the DUKW to hold the boxes clear of the cargo floor and to aid in loading and unloading. The heaviest boxes should be stowed in the forward end of the cargo space and all boxes should be lashed or blocked so that they will not shift.

f. Projectiles. 155-mm projectiles and similar ammunition are normally loaded in clusters of from 6 to 12 by means of hooks through the lifting eyes in the nose of the projectiles. When loaded in this manner, a considerable amount of small broken dunnage or nets should be used on the floor of the DUKW to prevent the projectiles from rolling. A heavy pallet, with sides approximately 24 inches high, may be used in place of the cluster of hooks. If the loaded pallets are stowed in the DUKW, only sufficient dunnage to keep the projectiles from damaging each other will be necessary.

g. Bombs. DUKW's are well suited for handling bombs. Much valuable information can be found on this subject by referring to FM 9–39. Additional instructions for loading various types of bombs are as follows (see par. 93h for unloading details):

(1) 100-pound bombs have a maximum weight of approximately 145 pounds and are packed complete in crates. No dunnage is necessary. They should be stowed lengthwise in the cargo space, eight abreast in three rows. For *difficult* conditions, the first row at the forward end should be tiered 2 deep allowing a total load of 32 bombs weighing approximately 4,650 pounds. For *favorable* conditions, the first row should be tiered 3 deep and the second row 2 deep, totaling 48 bombs, weighing approximately 7,000 pounds. For *ideal* conditions, first and second rows should be 3 deep and the third row 2 deep, a total load of 64 bombs weighing approximately 9,300 pounds.

(2) 250-pound bombs weigh 256 pounds each. For *difficult* conditions they should be stowed, without dunnage, lengthwise in the cargo space in 4 rows, 5 abreast in each row. This is a total load of 20 bombs weighing approximately 5,100 pounds. For *favorable* conditions, there should be a second layer of 4 bombs on both the first and second rows, a total of 28 bombs and approximately 7,150 pounds. For *ideal* conditions, all 4 rows should have a second layer of 4 bombs, giving a total of 36 bombs weighing approximately 9,200 pounds. Tail assemblies for these and all larger bombs are handled separately.

(3) 500-pound bombs have a maximum weight of 528 pounds. Dunnage should be used both to protect the floor and to prevent the first layer from rolling. The bombs

should be stowed lengthwise. For *difficult* conditions, there should be 2 rows of 4 bombs each, as far forward in the cargo space as possible. This gives a total load of 8 bombs weighing approximately 4,200 pounds. For *favorable* conditions, 3 additional bombs may be placed on top of each row, giving a total of 14 bombs weighing approximately 7,400 pounds. For *ideal* conditions, 4 extra bombs may be stowed lengthwise on the floor at the rear end of the cargo space. This gives a total load of 18 bombs weighing approximately 9,500 pounds.

(4) 1,000-pound bombs have a maximum weight of 1,041 pounds. Dunnage should be used both to protect the floor and to wedge the bombs in place. For *difficult* conditions, 1,000-pound bombs should be stowed 3 abreast, lengthwise, at the forward end of the cargo space. Two more bombs are loaded on top of the first 3, a total of 5 bombs weighing approximately 5,200 pounds. For *favorable* conditions, they should be stowed 3 abreast, lengthwise, at the forward end of the cargo space and 4 abreast, crosswise, at the after end of the cargo space, totaling 7 bombs weighing approximately 7,300 pounds. For *ideal* conditions, 2 more may be piled on top of the first 3, lengthwise, at the forward end, a total load of 9 bombs weighing approximately 9,350 pounds.

(5) 2,000-pound bombs have a maximum weight of approximately 2,093 pounds. Heavy dunnage must be used to protect the floor, to wedge the bombs securely, and especially to prevent them from rolling back when the DUKW climbs out at the water's edge. (See fig. 47.)

For *difficult* conditions, 2 bombs, weighing approximately 4,200 pounds, are stowed crosswise at the forward end of the cargo space.

For *favorable* conditions, 3 bombs, weighing approxi-

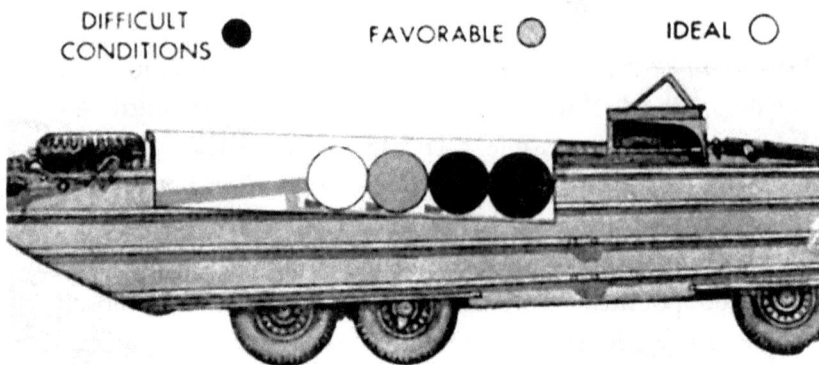

*Figure 47. Positioning of 2,000-pound bombs
under various conditions.*

mately 6,300 pounds, are stowed crosswise at the forward
end of the cargo space.

For *ideal* conditions, 4 bombs, weighing approximately
8,400 pounds, are stowed crosswise at the forward end of
the cargo space.

(6) 4,000-pound bombs weigh 4,152 pounds. Heavy,
dunnage must be used to protect the cargo floor when
they are loaded. Sufficient blocking must also be used

*Figure 48. Positioning of 4,000-pound bombs under DIFFICULT,
FAVORABLE, and IDEAL conditions.*

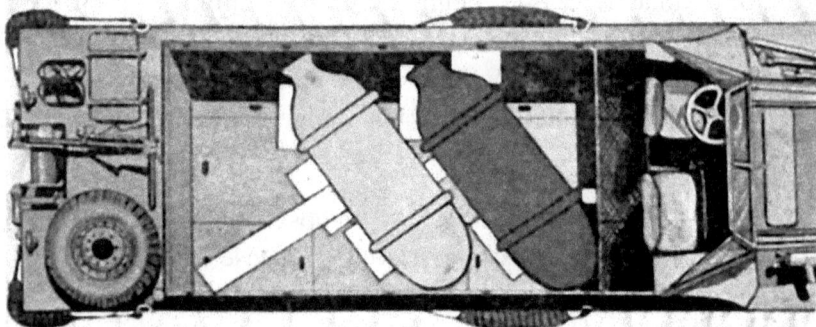

to cradle the bombs securely in the DUKW. Under *difficult* or *favorable* conditions, one bomb will be placed diagonally in the cargo space, with its nose in the right forward corner, and the tail end of the bomb against left coaming. To prevent the bomb from rolling or shifting, it must be heavily blocked. (See fig. 48.) The tail assembly may be carried in the stern. Under *ideal* conditions, a second bomb should be placed parallel to and alongside the first bomb. To prevent the bombs from rolling toward the stern, they must be heavily blocked. The two tail assemblies may be carried in the stern.

91. CARGO HANDLING DUTIES OF ASSISTANT DRIVER. a. After hooking in the mooring hook, the assistant driver should stand on the stern deck, ready to guide the draft to its proper position in the DUKW as it is lowered from the ship. The driver will assist from the *forward* end. (See fig. 42.)

b. After unhooking the first load, either the driver or the assistant driver will hook empty nets, pallets, slings, etc., from previous drafts to the cargo fall so they can be lifted aboard the ship for future use.

c. After the DUKW has been loaded and has pulled away from the ship, the assistant driver will secure the load, blocking and lashing any cargo that might shift, and shifting cargo where possible to trim the DUKW for its trip to shore.

d. As the dump is approached, the load should be arranged to speed unloading. Slings should be adjusted and shortened as necessary and temporary lashings cleared away.

e. On the return trip to the ship, the assistant driver will clean the bilges and cargo space, and perform "during operation" maintenance. Pallets, nets, and slings should be moved to clear the space in the DUKW where the first draft will be landed from the ship.

92. UNLOADING AIDS. a. General. To speed unloading it is important to use the best available methods, which will vary with each situation. The DUKW company must be prepared to operate efficiently with numerous alternatives, the most common of which are outlined below.

b. Crane. The crane, which can handle any cargo carried by the DUKW, is the fastest and most useful aid in unloading. It also saves manpower. (See fig. 49.)

Figure 49. The crane is the fastest aid in unloading.

However, there will seldom be enough cranes available to handle all the jobs required.

c. A-frame. The **A**-frame, which can be attached to the DUKW, will perform most of the functions of a crane. (See fig. 68.) It is, in effect, a crane that uses the winch and cable of the DUKW for hoisting. The usefulness of the **A**-frame is limited because of the length of time needed in unloading by this means and also because it ties up another DUKW. **A**-frames are usually furnished with one DUKW in five. Details of the **A**-frame and instructions in its use are as follows:

(1) All DUKW's are fitted with the necessary brackets for mounting the **A**-frame. Pins attach each leg of the **A**-frame to brackets on the stern and two stay wires lead from the ears at the head of the frame to the forward lifting eyes of the DUKW. The winch cable is led over the sheave in the head of the frame. A stiff-leg connects the frame to the spare tire and wheel on the DUKW to prevent jackknifing of the **A**-frame when maneuvering without a load. An **A**-frame that is allowed to jackknife will become bent and be unable to support a heavy load.

(2) DUKW's equipped with **A**-frames should be operated on reasonably level ground. The **A**-frame should not be used afloat without taking special precautions to stabilize the DUKW. (See figs. 65 and 66.)

(3) To unload cargo, the **A**-frame DUKW backs up at right angles to the DUKW carrying the cargo. The assistant driver of the loaded DUKW will hook on the load and signal the driver of the **A**-frame DUKW to guide the hoisting of the load and the subsequent maneuvering of the DUKW. The two basic operating plans that may be followed for unloading a DUKW by using an **A**-frame are shown in figure 50. The one that is best suited to the dump being worked should be used.

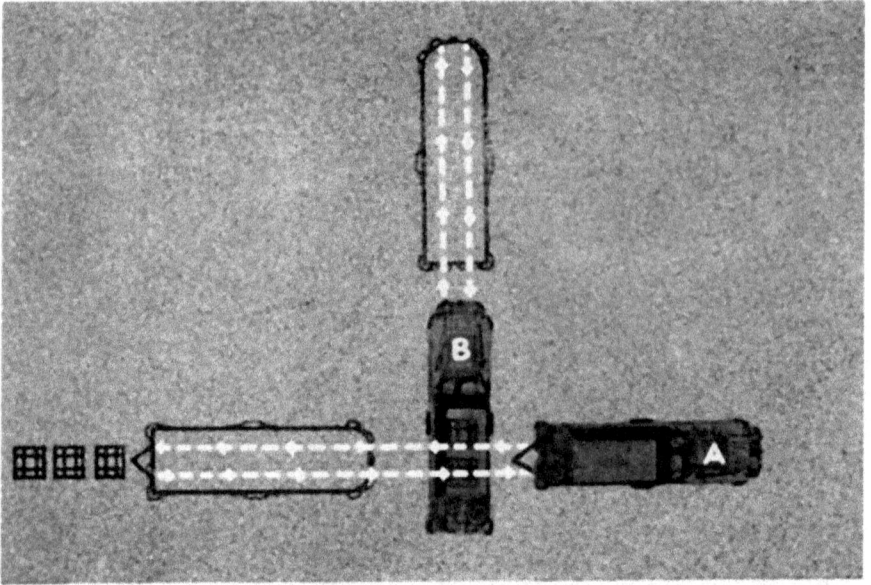

Operating plan 1. a. The **A**-frame DUKW (*A*) backs up at right angles to the cargo carrying DUKW designated as *B* and lifts the cargo.

b. DUKW *B* moves forward out of the way.

c. DUKW *A* moves backwards and deposits draft on the ground, then returns to former position.

d. DUKW *B* returns to former position and procedure is repeated until cargo is unloaded.

*Figure 50. Basic operating plans for unloading a DUKW when using an **A**-frame.*

(4) When a load is lifted with an **A**-frame, the winch power take-off lever in the cab will be placed in neutral before moving the DUKW. If this is not done, the load will be lifted higher and may cause damage. Before the DUKW is moved any distance, the load should be lowered to a point near the ground and steadied by helpers on the ground or lashed to the stern pintle to prevent swinging. The winch drag brake must be kept properly adjusted or the load being carried may drop without warning.

(5) The **A**-frame is not suited for unloading if a long lift is necessary because the top of the cargo net will reach

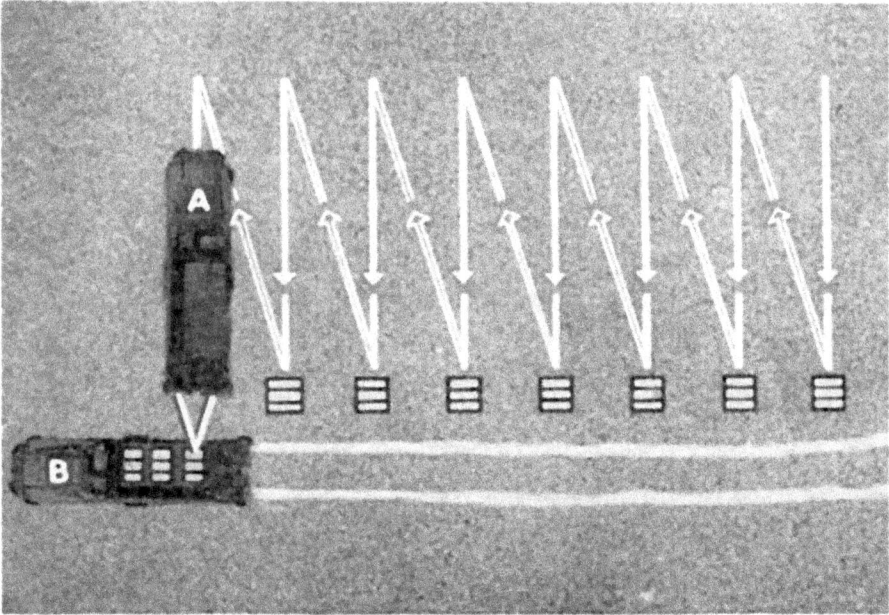

Operating plan 2. *a.* The **A**-frame DUKW (*A*) backs up at right angles to the cargo carrying DUKW designated as *B* and lifts the cargo.

b. DUKW *A* moves forward just clear of DUKW *B* and deposits the draft on the ground.

c. DUKW *A* continues forward and then backs up to the next unloading position and the procedure is repeated.

d. DUKW *B* moves forward only as necessary.

Figure 50. Basic operating plans for unloading a DUKW when using an **A**-*frame—*Continued.

the apex of the frame before the bottom of the draft has cleared the DUKW coaming. The **A**-frame can be used to handle cargo nets if barrel hooks or spider hooks are used, hooking them in down low on the sides of the net. (See fig. 51.)

(6) The maximum weight that should be lifted with an **A** frame is 5,000 pounds. To lift this load, 1,000 pounds of cargo or personnel must be placed on the forward deck for ballast. Approximately 4,000 pounds is the limit that can be lifted with an **A**-frame without ballast.

(7) Temporary **A**-frames can be made in the field with

3-inch pipe, 5- x 5-inch timbers or their equivalent. Each leg should be 12 feet 5 inches long, a length which will permit stowing them in the cargo space when not in use. Higher frames are dangerous from the standpoint of stability. The snatch blocks which come with the DUKW can be used at the apex of temporary **A**-frames. The **A**-frame should overhang the stern of the DUKW by 4 feet.

Figure 51. Spider hooks should be used to remove net loads with an **A**-frame.

Figure 52. Ten-foot stationary hog trough.

d. Hog troughs. Where cranes or **A**-frames are not available, and where there are many packages to be unloaded by hand, hog troughs will be found useful. These are made by nailing two 2- x 12-inch planks to form a **V** as shown in figures 52 and 53. There are two types of hog troughs:

(1) A trough to be used while the DUKW is stationary. This is 10 feet long and has a block at the deck and a block at the ground, as shown in figure 52.

(2) *A trough 6 feet long to be used while the DUKW is moving.* The upper end of the trough is fitted with 1/4- x 1 1/2-inch strap hooks to secure it over the coaming. It has a 3-foot leg to support the extended portion of the trough at an angle of 45° so that the end is about a foot clear of the ground. With this type of trough the DUKW can be driven along slowly as packages are unloaded. This affords dispersion and eliminates the necessity of rehandling the cargo at the base of the trough. (See fig. 53.)

Figure 53. Cargo can be unloaded while the DUKW is moving with these 6-foot hog troughs.

Figure 54. Barrel skids for hand unloading of 55-gallon drums.

e. Barrel skids. Barrel skids (fig. 54), made of two 4 inches x 4 inches x 10 feet long boards and secured 14 inches between inner edges, provide a useful method for unloading 55-gallon drums by hand. The proper procedure for using this barrel skid is explained in paragraph 93e(2).

f. Plank skids. Plank skids, measuring 2 inches x 12 inches x 12 feet, are very simple to employ, as illustrated in figure 55.

Figure 55. Plank skids measuring 2 inches x 12 inches x 12 feet are a simple expedient.

g. Roller runways. Roller runways can be used in a manner similar to plank skids. They may also be used on the ground to distribute material unloaded by means of hog troughs or plank skids.

h. Hand unloading. Where enough men are available, and the packages are not too heavy, rapid unloading can be accomplished by hand unloading.

Figure 56. Platform made of cargo helps when handling heavy boxes.

i. Platform. When packages are heavy, they may be piled evenly on the ground to form an unloading platform, as shown in figure 56, which will provide a more convenient working height.

93. NORMAL UNLOADING PROCEDURES. a. General. Where possible, mixed loads that involve going to more than one dump and loads that involve unloading with several types of equipment in any one dump should be avoided.

b. Nets. When discharging nets from a DUKW with **A**-frames or short cranes, it will be found that the height or drift of this equipment will be insufficient. This can be overcome by using six sets of barrel hooks or a six-legged spider bridle with hooks that can engage the mesh of the nets. (See fig. 51.)

c. Pallets. When unloaded pallets have been carried into the dump, they should be lifted out of the DUKW by using a crane or **A**-frame fitted with suitable gear. Where a crane or an **A**-frame is not available, the pallets should be unloaded by using hog troughs, plank slides, roller runways, or by hand.

d. Bound pallets. Bound pallets should be unloaded intact by crane or **A**-frame and not opened until time of actual use. If this method is not practicable, the straps on the pallets must be broken and the cargo discharged by hog troughs, plank skids, roller runways, or by hand.

e. Fuel drums. (1) The quickest way to unload 55-gallon drums is by crane or **A**-frame, using six sets of barrel hooks, as shown in figure 45, or a cargo net, as shown in figure 51. Rope slings also may be used, although it is safe to lift only three drums at one time in this manner.

(2) If mechanical equipment is not available, barrel skids constructed as suggested in paragraph 92e will provide a satisfactory method for the unloading of drums by hand. The skid is placed against the side of the cargo space and the drums are slid over the coaming and down the skid to the ground where they can be rolled into storage piles. Drums on the bottom of the cargo floor must be placed on end about 2 feet from the coaming and then leaned against the coaming. Next, with two men lifting the bottom end, the drum is slid up over the coaming and down the barrel skid as shown in figure 57.

f. Boxed ordnance parts. Boxed ordnance parts must be unloaded by crane or **A**-frame with rope or wire slings.

g. Projectiles. Projectiles are most easily unloaded by a crane or **A**-frame either in pallet or with a cluster of 6 to 12 hooks. When a crane or an **A**-frame is not available, the projectiles should be unloaded by hand with the aid of a hog trough.

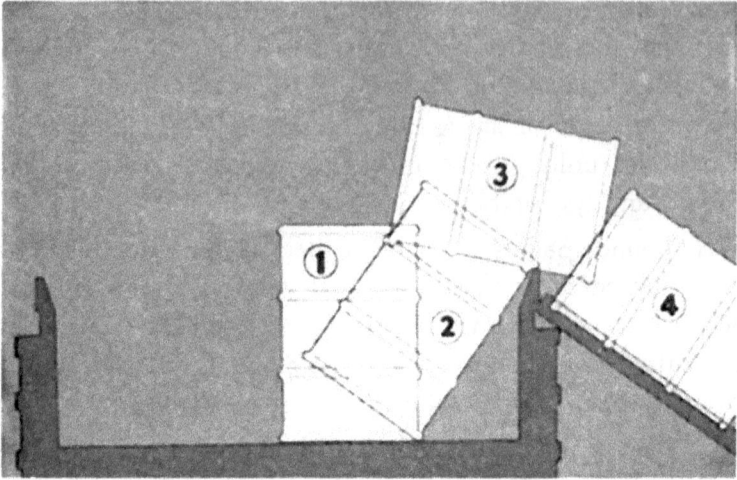

1. Stand drum on end at distance from side that will permit upper ring to come just above coaming when tipped to position 2.
2. Two men lift up and outward to position 3.
3. Guide drum over center of skids and slide it over coaming onto skids.
4. Let drum slide down skids to ground.

Figure 57. Procedure for hand unloading of fuel drums with barrel skid. Work close to rear end where cargo space is shallowest.

h. Bomb unloading. The method used in unloading bombs depends upon the type of bombs being carried. Various unloading methods are outlined below:

(1) 100-pound bombs should be unloaded by hand using a hog trough, or, when loaded on pallets, by the use of an **A**-frame or crane if available.

(2) 250-pound bombs should be unloaded by an **A**-frame or crane when loaded on pallets, or by a cluster of hooks or similar gear if individually loaded. When the bombs are unloaded by hand, hog troughs should be used.

(3) 500-pound bombs should be unloaded by **A**-frame or crane with bridles slung to pick up four bombs at a time

or with suitable pallets. These bombs can also be rolled up on skids over the side of the cargo space and dropped onto soft ground *if mechanical aids are not available.*

(4) 1,000-pound bombs should be unloaded by **A**-frames or cranes, with bridle or slings to handle two bombs at a time. In an emergency, where mechanical aids are not available, bombs may be rolled up on skids over the side and dropped onto soft ground.

(5) Unloading 2,000- and 4,000-pound bombs will be accomplished with **A**-frame or crane fitted with suitable bridle or slings. (See fig. 58.)

Figure 58. Bridle for 2,000-pound bombs.

94. DISCHARGING OF CARGO FROM LST (2) INTO DUKW's.

a. General. If the LST (2) is beached and there is no appreciable water gap between the beach and the LST, land vehicles should be used for unloading. If, however, there is a water gap between the beach and the LST, or if the LST is anchored off shore, DUKW's will be found useful in discharging cargo.

b. Entering. The DUKW's will enter the tank deck of the LST by going in over the ramp, bow first.

c. Unloading. Five DUKW's can be loaded simultaneously with stores from the upper deck of this type of landing ship in the following manner (see fig. 59):

(1) Two DUKW's should be driven parallel on the tank deck to a position just astern of the after hatch. These DUKW's can be loaded with the aid of chutes extending down from the top deck, the cargo being fed down the chutes to the DUKW's.

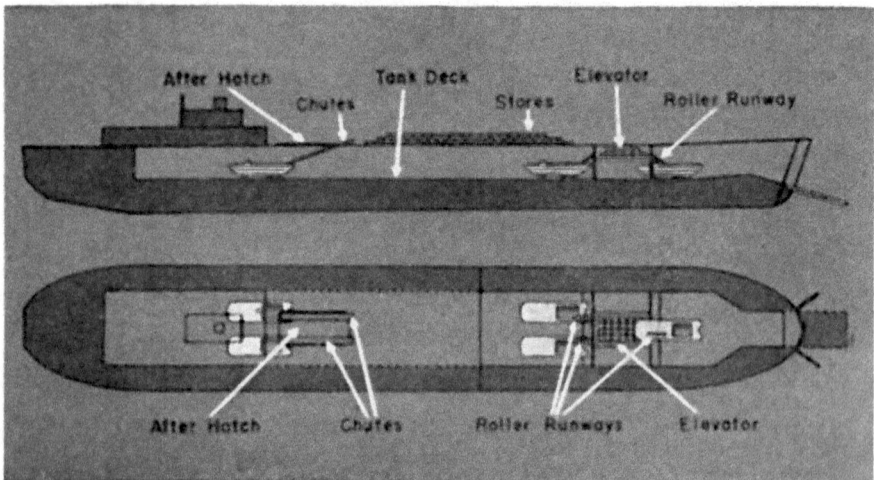

Figure 59. Discharging of cargo from LST's (2) into DUKW's.

(2) Two DUKW's should be driven to a position just astern of the elevator and another DUKW driven just forward of the elevator. The elevator, which will have been loaded with cargo approximating the capacity of the three DUKW's, should be lowered to a position slightly above the DUKW's. Cargo can then be passed from the elevator to the DUKW's with the aid of chutes or roller runways or by hand.

d. Leaving. The loaded DUKW's will be backed out stern first, going *slowly* down the ramp. If the DUKW is heavily loaded, the rear canvas or plywood closure must be set up and the tarpaulin rigged over the rear half of the cargo space preparatory to backing down the ramp.

SECTION XIII

VISUAL SIGNALING

---

95. GENERAL. DUKW operations require the use of both conventional arm signals, employed in road driving on land, and special signals to be used when the DUKW is on the water.

96. LAND SIGNALS. On the road the driver will use and be governed by the arm signals which are found in TM 10–460.

97. WATER SIGNALS. **a. General.** On the water there are several means of communication that may be used. These generally are based on the International Morse Code or Semaphore Code.

b. International Morse Code. International Morse Code signaling is carried out by light, sound, or wigwag. The whole range of alphabetical and numerical symbols is based on two elements called a "dot" and a "dash." See fig. 60 for International Morse Code.

(1) *Light.* (*a*) In this system of visual signaling, the dot and dash are transmitted by intermittent beams of light. This may be accomplished by switching the lights on and off at the correct intervals, by inserting a blackout screen in front of the lights, or by use of blinkers. A mirror may also be used for this purpose.

(*b*) In transmitting messages by light by International Morse Code, a "dot" is regarded as one unit of duration

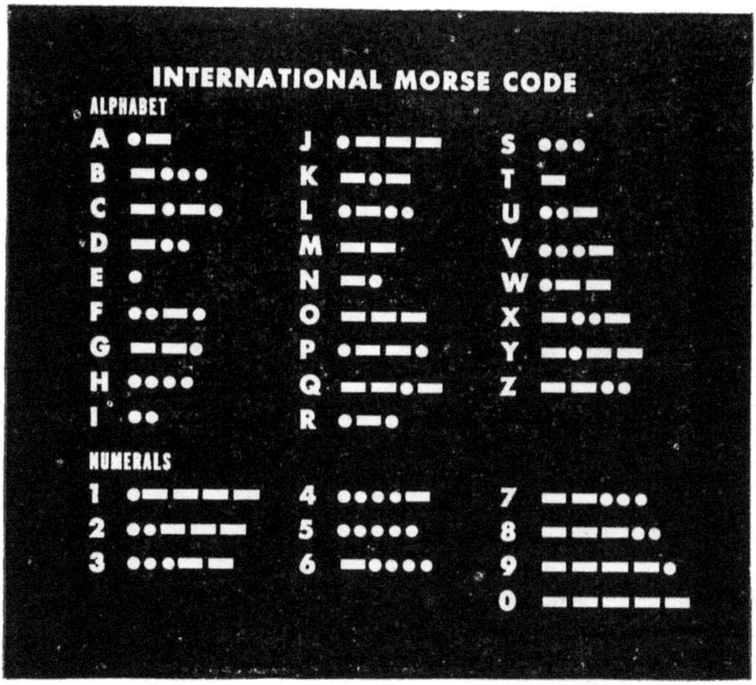

INTERNATIONAL MORSE CODE

ALPHABET

A ●■	J ●■■■	S ●●●
B ■●●●	K ■●■	T ■
C ■●■●	L ●■●●	U ●●■
D ■●●	M ■■	V ●●●■
E ●	N ■●	W ●■■
F ●●■●	O ■■■	X ■●●■
G ■■●	P ●■■●	Y ■●■■
H ●●●●	Q ■■●■	Z ■■●●
I ●●	R ●■●	

NUMERALS

1 ●■■■■	4 ●●●●■	7 ■■●●●
2 ●●■■■	5 ●●●●●	8 ■■■●●
3 ●●●■■	6 ■●●●●	9 ■■■■●
		0 ■■■■■

Figure 60.

and a "dash" as equivalent to three units. This length of units must be adhered to regardless of the speed of transmission. A space of two units should be allowed between characters and a space of three units between words.

(2) *Sound.* International Morse Code messages also may be transmitted by means of intermittent sounds. The DUKW's horn may be used for this purpose. The same principles as outlined in (*b*) above should be followed.

(3) *Wigwag.* (*a*) Wigwag is a method of transmitting a message in which the dots and dashes are made by the swing of the arm. (See fig. 61.)

(*b*) The man sending the message should hold a flag, handkerchief, or cap overhead in his right hand and face

121

Figure 61.

WIGWAG

the receiver. A dot is then made by swinging the arm downward and sidewise to the sender's right through an arc of 90° and immediately bringing the arm back to the starting point overhead. A dash is made by a similar motion to the sender's left. A short pause at the overhead position indicates spacing between the individual characters. The end of a word or group of words is indicated by dipping the flag to the front of the operator and the end of a message by two such motions.

(*c*) For signaling at night, a handlamp or flashlight may be substituted for the flag, handkerchief, or cap. In this case, the swing is made upward to a horizontal position from a starting point with the arm hanging down naturally, instead of downward from a starting point overhead.

(*d*) The ATTENTION sign is made by waving the flag several times through a semicircle (180°) overhead, or by waving the lantern through 180° across the knees.

c. Semaphore. (1) Semaphore is a means of visual signaling by the use of a flag in each hand. The system provides for its own alphabet, numerals and other symbols. (See fig. 62.) In semaphoring, the arms are placed at

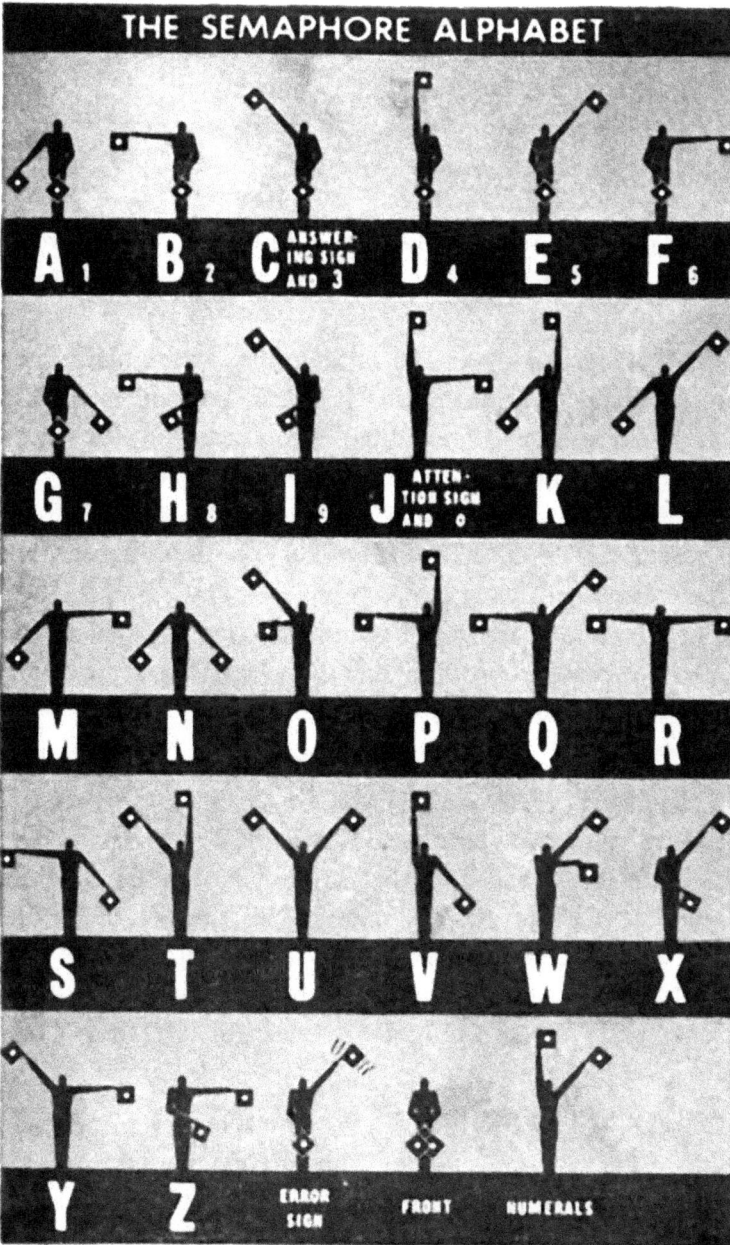

THE SEMAPHORE ALPHABET

Figure 62.

the exact positions indicated, a distinct pause is made at each position, and the arms are moved from one position to another by the shortest possible route.

(2) The signs provided for in semaphore in addition to the alphabet and numerals are:

(a) *Answering sign*. The letter "C" is used by the receiver to acknowledge the correct reception of each word. If this letter signal is not made, the sender must repeat the word.

(b) *Attention sign*. To indicate that a semaphore message is about to be made the "J" letter position is used.

(c) *Error sign*. Should an error occur in the transmission of a message, it is indicated by the successive signaling of the error sign.

(d) *Front sign*. The front sign is used before and after each word.

(e) *Numeral sign*. The numeral sign is used before and after each group of numerals, when they are made as numbers and not spelled out.

98. SIGNAL FOR ASSISTANCE AT SEA. There will be occasions when the driver of a DUKW afloat will need the assistance of some other craft. In the daytime, he should lash a life jacket to one end of the boat hook and hold the boat hook in the air in a vertical position. At night if the tactical situation permits, he should flash the usual "SOS" signal by lights in International Morse Code.

99. MOORING SIGNALS. a. The company commander should be sure that each of his drivers understands the system of signals to be used for calling a vehicle to a specified position alongside a ship which is to be unloaded by DUKW's. He should first assign a number or identifying mark to each vehicle, and each driver should be thoroughly familiar with the number assigned to him. Each mooring station at ship's side will also be assigned a number, stations being numbered from bow to stern, those on the right side having odd numbers, those on the left being numbered evenly. Thus a Liberty ship with 5 cargo hatches will have stations 1, 3, 5, 7, and 9 on the right (starboard) side; and stations 2, 4, 6, 8, and 10 on the left (port) side.

b. When a vehicle is off the stern of the ship awaiting orders to proceed to one of the mooring stations, the assistant driver must watch for his signal. During daylight hours, the DUKW number should be displayed on a large white card held up on the stern deck, and under it will be another card with the mooring station number. At night, the DUKW number and mooring station number may be flashed by lights.

UNIT OPERATIONS AND CONTROL

100. CONTROL. By their very nature, the operations of an amphibian truck company, with individual vehicles continuously running from shipside offshore to inland dumps, present a complex problem of control. Nevertheless, the amount of cargo which can be moved by a group of DUKW's depends to a great extent on efficient control.

101. DUKW CONTROL SYSTEM. To insure necessary control, the following system, built around a DUKW control center, should be established on the shore. The control center is an operational headquarters. From it, all phases of the operation are coordinated by personnel who report to the control center but operate at key points on board ships being worked, in dumps being worked, and in the dispersal area ashore. (See fig. 63.)

102. DUKW CONTROL CENTER. As stated above, the DUKW control center is the headquarters for all DUKW operational activities. The center should be located near the point where the DUKW's enter and leave the water and, if possible, in sight of the ships being worked.

103. DUKW CONTROL OFFICER. One officer of the amphibian truck company is designated as DUKW control officer and is assisted by such other officers and noncommis-

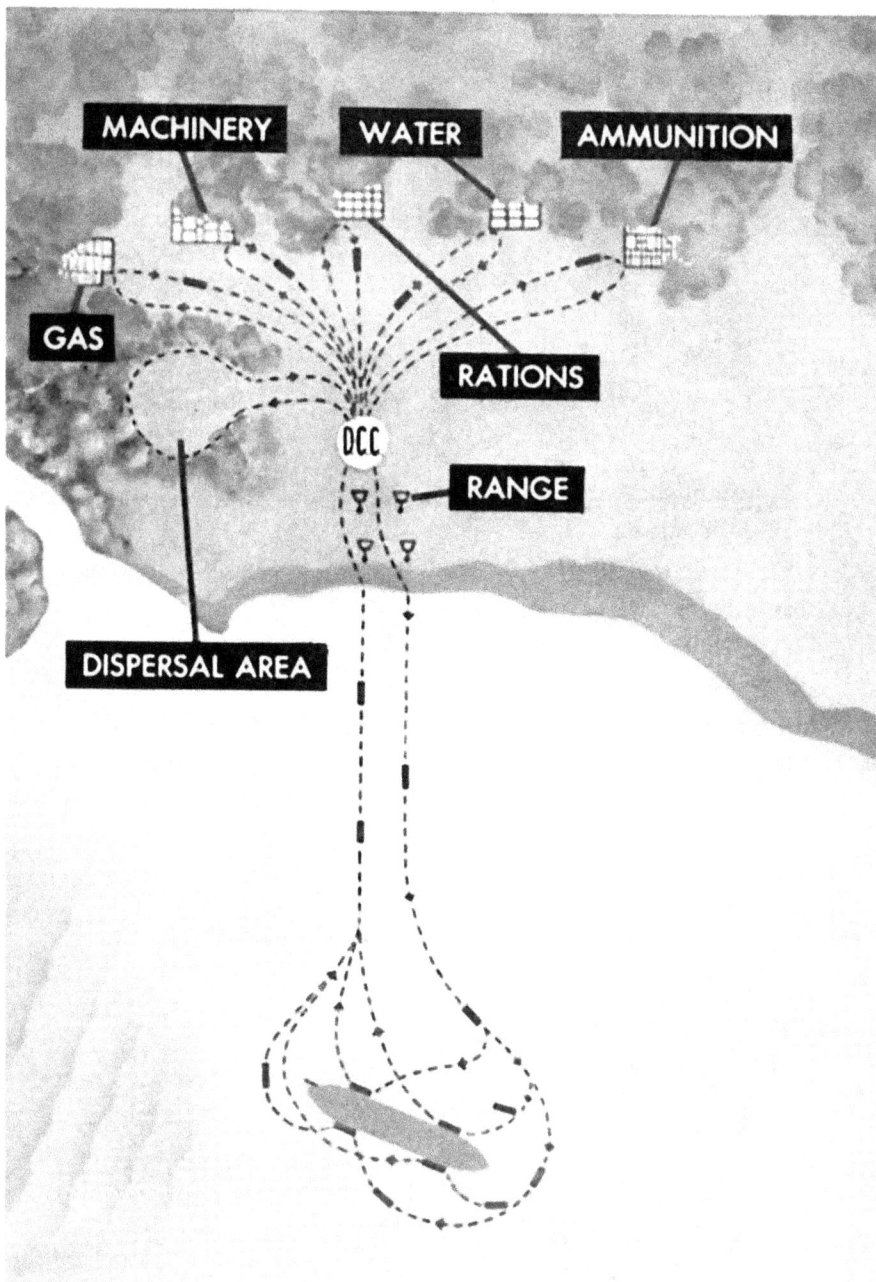

Figure 63. Typical DUKW operations, including
DUKW control center.

127

sioned officers as are available. Communication facilities, including the unit's visual signalmen, will be at his disposal when desired. Information pertinent to the effective control of operational activities should be communicated to the DUKW control officer. This information includes such data as the number of DUKW's available, the ships to be unloaded, the nature of cargo to be handled, and the location of the dumps where the cargo is to be unloaded. Duties of the DUKW control officer include—

a. Establishment of communications system between all DUKW control points.

b. Assignment of available DUKW's to specific hatch positions of ships to be unloaded.

c. Maintenance of an efficient flow of DUKW's to and from ships during unloading activities.

d. Maintenance of a continuous check on DUKW movements in order to be able to advise higher authority of the number of vehicles in operation.

e. Maintenance of records of cargo passing through DUKW control center for purposes of advising higher authority.

f. Routing of loaded DUKW's to specific beachhead or inland dumps.

g. Arranging for beach matting where necessary.

h. Keeping landing place and tracks across beach clear of wreckage and objects which might puncture tires.

i. Marking the DUKW channels, and maintenance of range markers for day or night and route markers between beach and dumps.

j. Making decisions affecting DUKW operations arising from weather conditions.

k. Maintenance of liasion with beach master, port authorities, and similar officials.

104. LOADING CONTROL. A commissioned or noncommissioned officer representing the DUKW control officer should be aboard each ship at the time of unloading. This officer may be referred to as the loading control officer. His duties include—

a. Maintenance of communications with the DUKW control center.

b. Maintenance of liasion between the DUKW control center and the master of the ship being unloaded.

c. Recommending to the master of the ship any actions which may help speed up the unloading, such as the making of a lee or the shifting of anchorage in order to shorten the ship-to-shore distance.

d. Maintenance of the proper rigging and condition of the mooring equipment used by the DUKW's.

e. Supervision of the coming alongside and mooring of the DUKW.

f. Supervision of loading operations by DUKW company personnel.

g. Determining the payload of DUKW's depending on conditions.

h. Advising the DUKW driver of the nature and approximate weight of his cargo.

i. Advising the DUKW control center, at the appropriate time, of the amount of cargo and type to be unloaded, and other pertinent information.

105. DUMP CONTROL. The DUKW control center exercises whatever control is necessary to speed unloading of cargo from the DUKW's at the dump. The officer or noncommissioned officer assigned to this operation may be referred to as the dump control officer. His duties include—

a. Maintenance of communications with the DUKW control center.

b. Reconnaissance of dump area locations for the various classes of supplies and the routes to and from the DUKW control center.

c. Directing the efficient dispersal of cargo and DUKW's.

d. Seeing that all natural cover and concealment available are made use of, and making provision for artificial cover as needed.

e. Supervision of the proper unloading of cargo and arranging for available cranes, **A**-frames, and other unloading aids.

106. DISPERSAL CONTROL. The DUKW control center exercises whatever control is necessary to coordinate activities at the dispersal area. This area is used as a parking place for DUKW's that are either awaiting orders to proceed on a mission, or are scheduled for preventive maintenance operations. The officer or noncommissioned officer assigned to this area may be referred to as the dispersal control officer. His duties include—

a. Maintenance of communications with the DUKW control center.

b. The effective dispersal of vehicles.

c. Camouflage of vehicles and area.

d. With the assistance of the maintenance officer, providing facilities for operational maintenance necessary in this area.

e. Arranging for messing facilities, first-aid stations, and latrines.

107. AROUND-THE-CLOCK OPERATIONS. When engaged in unloading or other operations which cannot be completed in one 12-hour working day, the personnel of the amphibian truck company should be organized into two shifts and the work divided between them. Each shift should be organized with an administrative, a control, a maintenance, and an operations unit so that it can operate independently of the other.

108. DOUBLE SHIFT. When operations are on a double shift basis, the duties of the various units are as follows:

a. Administration. The administrative unit arranges to have breakfast served to personnel in each shift before operations begin, and distributes lunches to the men for the mid-shift meal. These lunches may be supplemented, when possible, at the dispersal area. Supper is served immediately after each shift completes its tour of duty.

b. Control. All operations should be carried out under the control system outlined in paragraphs 100 to 106.

c. Maintenance. Regardless of the length of the operation, maintenance must be carried out as described in paragraph 23. Preventive maintenance operations, however, should be scheduled so that not more than nine DUKW's are out of operation at one time. Monthly and semiannual maintenance may be performed during the day on a one-shift basis. The maintenance unit should have sufficient personnel on duty during all shifts to take care of emergency repairs. Weekly preventive maintenance will be done either night or day by the driver and assistant driver, since their vehicle is scheduled for maintenance and not for operations. Driver's daily preventive maintenance will be performed by the driver as indicated in TM 9–802.

d. Operations. Half of the drivers and half of the assistant drivers should be assigned to each shift. The driver and his assistant should work in different shifts as operators of the same DUKW to which they are normally jointly assigned. This assures continuation of the definite responsibility that has been established for the maintenance of each vehicle and its equipment.

(1) In double shift operations, DUKW company personnel will normally be assisted by personnel from port companies or by other troops. This additional personnel should be utilized in one of the methods described below:

(a) When sufficient men are available, an assistant should be assigned to the operator of each DUKW.

(b) When it is not possible to assign a second man to each DUKW, special crews should be assigned at each loading position at shipside. These crews place and secure the load in each DUKW in turn as it comes alongside, transferring from one DUKW to the next by the use of a Jacob's ladder or by some other safe method.

(2) When it is not possible to get sufficient additional personnel to assign special crews to each loading position, the operator of one DUKW should help the operator of another. In order to do this, the second DUKW should tie up along side the first DUKW which is moored at shipside in the usual manner. This method is practical only in very calm water.

109. OPERATIONAL CALCULATIONS. a. The company commander may frequently be called upon to make the following estimates and calculations:

(1) The maximum number of DUKW's that can be efficiently operated at each unloading position at shipside.

(2) The number of hours that it will take to unload a ship.

(3) Determination of the "use factor."

b. A method of computing these estimates is outlined in the following paragraphs.

110. MAXIMUM NUMBER OF DUKW's THAT CAN BE EFFICIENTLY OPERATED AT EACH UNLOADING POSITION. a. Calculation.

It is necessary to estimate first the average time needed for a loaded DUKW to go from the ship to the beachhead or inland dump, be unloaded, and return to the ship. (This may be determined by a trial run, if necessary.) This figure is then divided by the estimated average time required to load the DUKW at shipside. The result equals the number of DUKW's needed.

b. Example. (60 ÷ 5 = 12).

(1) Time in minutes needed to make a round trip.. 60
(2) Divided by the average loading time at ship-
side in minutes 5
(3) The result equals the number of DUKW's
needed 12

111. TO ESTIMATE NUMBER OF HOURS NEEDED TO UNLOAD A SHIP.

The number of DUKW's used in computing this estimate must not exceed the number that are available, nor can it exceed the number that can be kept working efficiently at shipside. (See paragraph 110.)

a. Calculation. The total number of DUKW's to be used in unloading a ship is multiplied by the number of round trips a DUKW can make per hour from ship to dump, and that figure is multiplied by the average payload. The result is divided into the number of tons of freight to be unloaded from the ship. The result is the number of hours needed to unload the ship.

b. Example. (37 x 1 x 2.5 = 92.5; 8000 ÷ 92.5 = 86+).

(1) Number of DUKW's to be used.................. 37

(2) Multiplied by the number of round trips a
DUKW can make per hour.. 1
       Equals DUKW loads per hour...................... 37
(3) Multiplied by the average payload per
DUKW 2.5
       Equals tons carried per hour...................... 92.5
(4) 92.5 divided into the number of tons to be
unloaded 8,000
(5) Result equals the time needed to unload the
ship in hours... 86+.

112. "USE FACTOR." **a. General.** To insure that the
maximum possible use is being made of all vehicles, company commanders should maintain a constant check on
DUKW operating efficiency. Wasteful methods and prac-
tices can thus be detected in their early stages and cor-
rective action taken at once. A "Use Factor" has been de-
vised which will aid the commander in detecting inefficient
use of men and equipment. The "Use Factor" is the ratio
in percentage that the actual output bears to the potential
output. This ratio may be computed for any period of
time; however, periods of at least 2 weeks duration will
prove most useful. The procedure for calculating this factor is outlined below.

b. Determination of actual output. (1) *Calculation.*
From the company records, the total tonnage of cargo
moved during the period under consideration is determined. This figure is multiplied by the average distance
in miles (one way) that the cargo was hauled from ship to
dump. This product is divided by the number of days
under consideration. The answer gives the actual output in
ton miles per day.

(2) *Example.* (*a*) Total tonnage of cargo
moved from 5 June to 20 June.............................. 10,800

(*b*) Multiplied by the average distance in miles (one way) cargo was hauled from ship to dump.... 2.2

(*c*) Divided by the number of days under consideration 15

(*d*) Equals actual output in ton miles per day.. 1,584

c. Determination of potential output. (1) *Calculation.* The total number of hours the company can operate per day is divided by the average time, in hours, it should take one DUKW to make a round trip between the ship and the dump. This figure is multiplied by the number of DUKW's that should be available for operation. This product is multiplied by the average payload, in tons, that a DUKW should carry. This product is then multiplied by the most efficient trip distance in miles from ship to dump. The result equals the potential use factor of the company in ton miles per day.

(2) *Example.* (*a*) Hours the company can operate per day (24) divided by the time it should take one DUKW to make a round trip from ship to dump (1 hour[1]) 24

(*b*) This figure multiplied by the total number of DUKW's that should be available (37), equals number of DUKW trips per day............................ 888

(*c*) The product multiplied by the average payload in tons that a DUKW should carry (2.5), equals tons per day.. 2,220

(*d*) The product multiplied by the most efficient distance in miles[2] from ship to dump............. 2.25

(*e*) Result equals potential tons miles per day.... 4,995

---

[1]Round trip time in this example computed as follows:
    5 minutes wait for mooring space.
    5 minutes to load.

d. Determining of "Use Factor." (1) *Calculation.* The actual output is divided by the potential output. The result equals the "Use Factor" in percentage.

(2) *Example.* (*a*) Actual output in ton miles per day 1,584

(*b*) Divided by the potential ton miles per day 4,995

(c) Equals the "Use Factor".......................... 32%

e. Analysis of results. When, on the basis of these calculations, a commander discovers that the company's "Use Factor" is low, he should take action to correct it. The first step should be a careful analysis of all components of the "Use Factor" calculations. Analysis of the operation during the period used in the preceding examples showed the following:

(1) In the actual output example, 31 DUKW's were the average number used per day. This is below the 37 DUKW's which is normally the minimum available. The preventive maintenance of the company needed improving.

(2) In the actual output example, 2.2 tons was the average payload carried. Better loading at shipside would have increased this average to 2.5 tons.

(3) In the actual output example, 2.5 miles distance

---

10 minutes from ship to shore.
13 minutes from shore to dump, including dispatching.
10 minutes to unload, including slight delay.
10 minutes from dump to shore.
7 minutes from shore to ship.
60 minutes, or 1 hour.

[2]At location under consideration, the nearest safe anchorage is .75 mile offshore. The dumps had to be located on the average 1.5 miles inland.

from ship to dump is longer than when the ship and dumps are more advantageously located.

(4) In the actual output example, operations were carried on 16 hours per day. This is only two-thirds of the 24 hour maximum. If necessary, the company should have been organized on the basis of 24 hour operation.

SECTION XV

SPECIAL USES AND UNUSUAL CARGO

113. GENERAL. Because of its amphibious nature and design, the DUKW will be found useful in many operations other than those connected with the primary mission of the company. This chapter mentions some of the more common special operations in which a DUKW may be used.

114. REFLOATING A BROACHED BOAT. a. General. The DUKW may often be called upon to refloat a broached boat. The DUKW's that assist in this operation should be rigged with heavy supplementary fenders or mats around the bow for protection. The sooner the attempt is made to refloat the broached vessel, the easier the job will be.

b. When engine of the broached vessel will run. The assisting DUKW will approach the broached vessel from the beach and place its bow against the stern of the boat at a right angle. As the water floats the broached boat, the DUKW will push the boat back into the water far enough for its propeller to take hold. The boat can then move off under its own power.

c. When engine of broached vessel will not run. This operation requires two DUKW's. The first assisting DUKW pushes the stern of the broached boat out to sea. At the same time, the second assisting DUKW moves out across the stern of the stranded boat, to which it is attached by a tow line. (See fig. 64.)

Figure 64. *Helping a broached boat when its
engine will not run.*

②

Figure 64. Helping a broached boat when its engine will not run—Continued.

The DUKW on the beach pushes as the one out in the water pulls. Drivers should wait for the right instant, then use all available power. The DUKW's will make their greatest effort when the wash from the breaker is under the broached boat, giving her partial floatation. If the broached boat has positioned itself parallel to the beach, it may be easier, because of the shape of the boat, to push the bow off the beach instead of the stern.

115. BEACHING A BROACHED BOAT. In case the boat must be repaired before it can be floated, it is essential that it be safely beached high and dry as soon as possible. This can be done by rigging the winch cables of two DUKW's to the broached boat and pulling her up on the beach. Each DUKW should double rig its cable by attaching the snatch blocks to the boat, using all the cable from the winches except the last five turns. When the rigging is complete, the drivers of the two DUKW's should dig in with their wheels. One of the assistant drivers, who is on the ground in a position to watch the entire operation, signals the drivers when the power of the two winches of the DUKW's should be applied. Advantage should be taken of any surf action in this winching operation. The broached boat should be moved slowly up the beach. If the boat is flooded with water or sand, it should be pumped or otherwise cleaned out as soon as the outside water level is below the top of the boat. A boat that is full of water is too heavy to move far.

116. TO REFLOAT A BEACHED BOAT. As soon as a beached boat has been cleaned of sand and water and the repairs to it have been completed, it may be refloated. In order to do this, the winches of two DUKW's are rigged as outlined in paragraph 115 and the boat is pulled down to the water's edge. Boats can seldom be winched backward since the skeg, propeller, and rudder assembly will either be damaged or impede the movement of the boat. The boat should, therefore, be pulled forward in a wide circle to avoid breaking the skeg. When the DUKW's approach the water's edge, they may not have enough traction to pull the boat and other DUKW's or bulldozers should help push the boat into the water. A heavy boat that is in the water well offshore can help considerably by pulling on a line attached to the bow of the beached boat.

117. SALVAGE WORK. a. General. Because of its amphibious nature, power winch, **A**-frame, and towing ability, the DUKW may be used for improvised salvage work. Stalled equipment, stores that have been dropped

*Figure 65. Stabilizing an **A**-frame DUKW for use afloat by rigidly securing to another DUKW.*

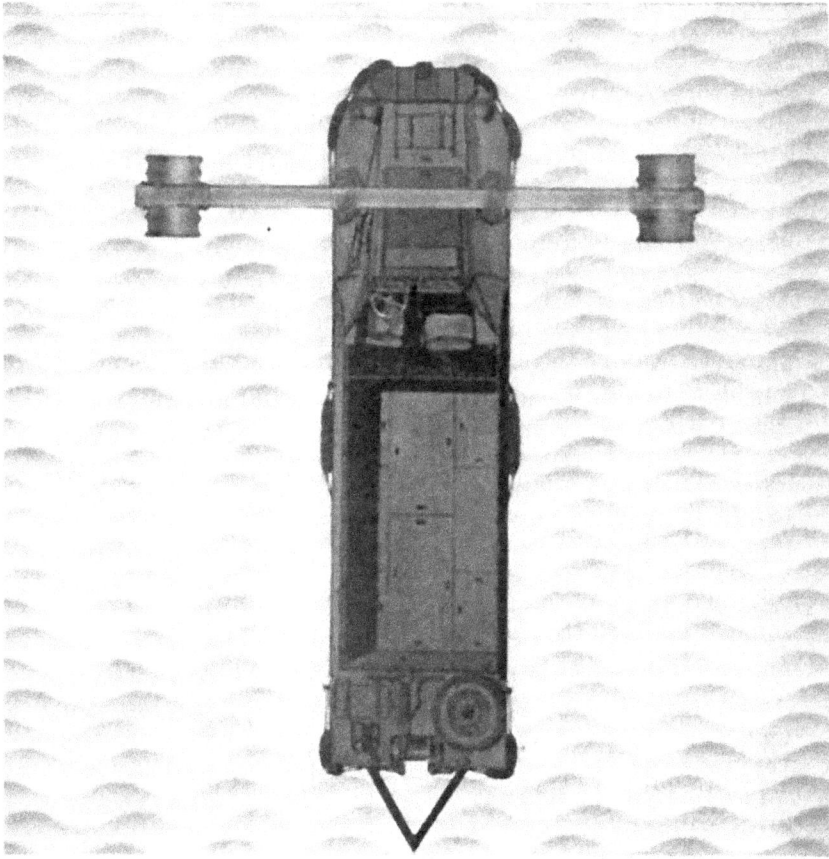

*Figure 66. Stabilizing an **A**-frame DUKW for use afloat by rigging two drums on a beam lashed to the lifting eyes.*

overboard, and beach obstruction may be salvaged or removed by towing or with the **A**-frame.

b. Deep water salvage and diving operations. Deep water salvage and diving operations can be conducted from DUKW's. The air compressor and tire pumping hose may be used with improvised diving equipment. In using the **A**-frame when fully afloat, it is necessary to stabilize the **A**-frame DUKW, as shown in figure 65 or figure 66.

Otherwise, the load on the apex of the **A**-frame, 12 feet above the water, would capsize the DUKW.

   c. Salvaging a sunken DUKW. If a DUKW has sunk in water that is not more than approximately 35 feet deep and the bottom is comparatively smooth, it may be salvaged by using two other DUKW's to tow it out. This operation should be performed when the water is comparatively calm. Generally, a sunken DUKW rests on its wheels. An effort should be made to determine the direction in which the sunken DUKW is heading. If the winch cables of the salvaging DUKW's cannot be attached by hand, it may be done as follows: The two assisting DUKW's tow their winch cables, joined together and dragging on the bottom between them, and approach the sunken DUKW from ahead. The cables should then be pulled under the front wheels of the sunken DUKW. When this is done, the cables will momentarily slacken. Shortening the cables and reversing the direction of the assisting DUKW's will make it possible for the cables to catch behind the front wheels of the sunken DUKW so that it can then be dragged to shore.

118. RECONNAISSANCE FOR BEACHING OPERA-TIONS.
The ease with which a DUKW can run from dry land into deep water affords a quick method of exploring, charting, and sounding shallow water approaches to a beach, and will prove of great help in selecting the best spot for beaching various types of craft and for other purposes.

119. EVACUATION OF CASUALTIES. a. General.
Twelve standard Army Medical Corps litters, each with a casualty, may be stowed in one DUKW. (See fig. 67.) Two attendants, exclusive of the driver and assistant driver,

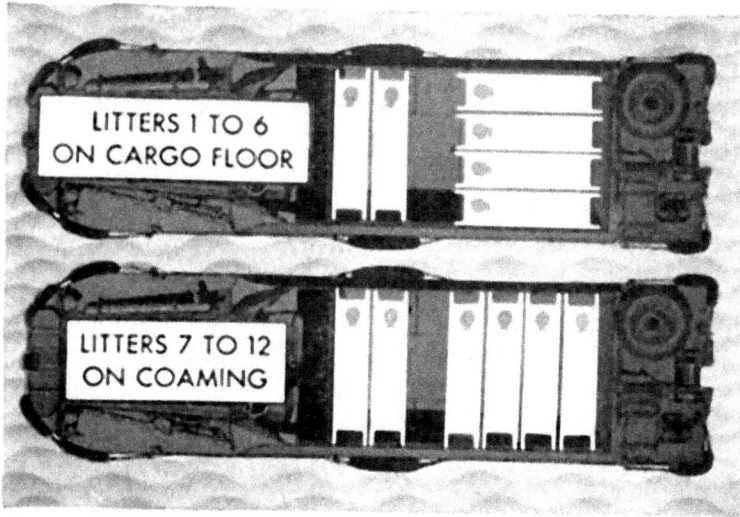

Figure 67. The DUKW will take 12 stretchers when properly loaded.

may also be carried. Once casualties are in the vehicle, they need not be handled again until the DUKW itself has been picked up at ship's side and deposited on deck. Once on deck, the men are transferred to special litters and carried to the sick bay.

b. Arrangement. The casualties on stretchers three, four, five and six should be laid with their heads toward the bow to enable the attendant to give them any necessary aid.

c. Tarpaulin. The tarpaulin should be rigged as usual if there is a possibility of spray, sun, or cold bothering the casualties.

d. Hoisting DUKW. If the ship's boom and cargo gear are sufficiently strong, the DUKW may be hoisted aboard and deposited on deck by the use of a 4-legged sling (16-foot legs). The hooks at the ends of the sling are made fast to the four lifting eyes.

e. Hoisting litters. If the ship's cargo gear is not sufficiently strong to hoist the DUKW aboard, the casualties may be hoisted aboard individually in nets, on pallets, or by any other suitable means.

120. FRESH WATER SUPPLY. A standard 600-gallon unit (section from a Navy T–6 Quonsette barge or similar unit) can be secured at the center of the forward end of a DUKW cargo compartment. Thus equipped, the DUKW can bring fresh water from a ship offshore to a desired spot inland.

121. UNUSUAL TYPES OF CARGO. a. Truck, ¼-ton, 4 x 4. A truck, ¼-ton, 4 x 4, can be placed in the DUKW cargo space with no special preliminary preparation. Dunnage should be used to distribute the wheel load over the cargo floor. The brakes should be set and the wheels lashed or chocked to prevent rolling. The truck is unloaded ashore with slings by the use of a crane or a DUKW fitted with an A-frame.

b. 37-mm antitank guns. Two 37-mm antitank guns can be carried in the DUKW cargo space, one at the forward end, barrel forward, the second at the aft end, barrel pointed toward the stern. Wheels should be chocked or lashed to prevent rolling. The guns can be unloaded ashore by crane or A-frame.

c. 105-mm howitzer. One 105-mm howitzer can be carried in a DUKW. Dunnage will be placed under the howitzer's wheels which should be placed in the forward corners of the cargo space. Some early model DUKW's, serial number 353–1 to 353–1005, must have the forward end of the coaming guard bent out slightly to clear the tires and hubs of the howitzer. A cargo net makes a good bumper to protect the rear deck and coaming from the

146

trail spade. Brakes should be set, the wheels chocked, and the howitzer lashed down to prevent rolling. The howitzer can be unloaded ashore by crane or A-frame.

Figure 68. 105-mm howitzer can be carried in a DUKW.

d. **Cargo with high center of gravity.** Occasionally the DUKW will be called upon to carry a load with a comparatively high center of gravity. The stability of the DUKW while afloat depends on its center of gravity which in turn is affected by the height and position of its load. The graph shown in figure 69 indicates the safety limits of the height above the cargo floor of the center of gravity of loads of various weights. The heavier loads should generally be placed as far forward as possible in the cargo space. (See par. 89.)

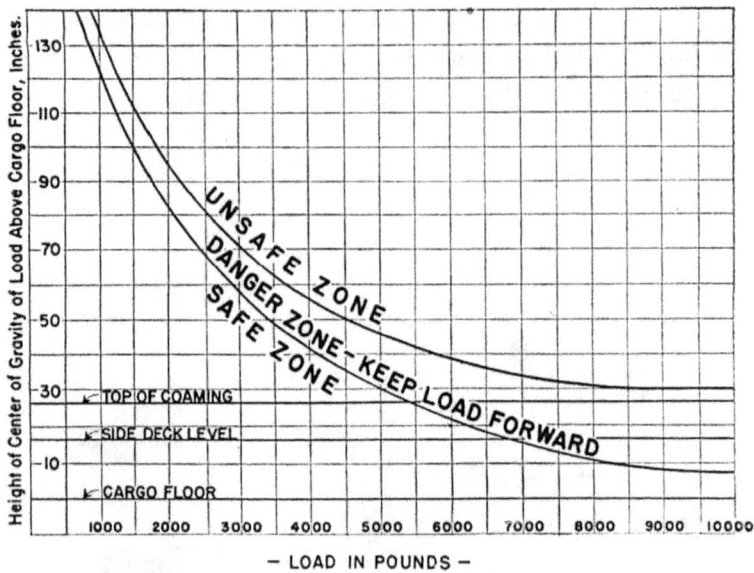

Figure 69. Center of gravity safety limits.

SECTION XVI

FIELD OPERATIONS

122. GENERAL. The amphibian truck company may be employed as a company, which is preferable, or platoons, sections, or squads may be assigned to separate missions. When units of the company are employed separately, they must include necessary maintenance personnel and equipment. The maintenance personnel and equipment of the company, or unit when operating independently, must be supplemented by ordnance shops for major repairs and replacement of major units.

123. ANTIAIRCRAFT DEFENSE. a. Certain trucks in the amphibian truck company are equipped with .50-caliber machine guns mounted for all around traverse and capable of high-angle fire. (See fig. 70.)

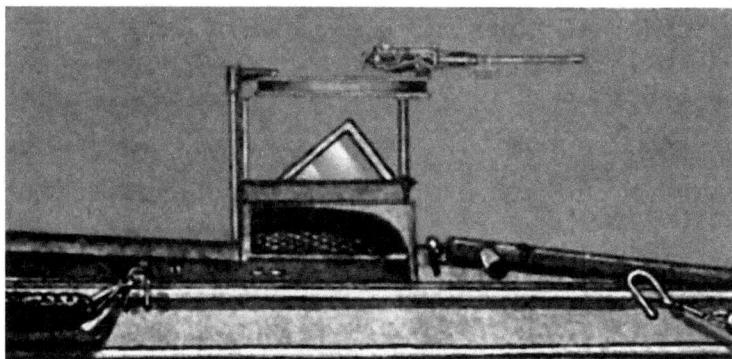

Figure 70. A .50-caliber machine gun on scarf ring mounted on a DUKW.

A sufficient supply of belt-loaded ammunition should be kept at all times in the vehicles mounting these guns. Although these weapons are effective against low-flying aircraft, the small number of weapons provided and the fact that the motion of the DUKW in the water makes accurate laying of the gun difficult require that the chief antiaircraft defense of the unit be provided by friendly aircraft and by ships' and shore antiaircraft artillery. Therefore, unless attack is imminent, hostile aircraft should not be fired upon since such fire mav only disclose the position of the DUKW.

b. Other antiaircraft defensive actions to be taken are—

(1) Camouflage or concealment on land.

(2) Dispersion on land or on water.

(3) Evasive action on land or on water.

(4) Slit trenches and foxholes on land.

(5) Smoke screens on water or on land.

c. For approved methods of camouflage see—

FM 5–20, Camouflage, Basic Principles

FM 5–20B, Camouflage of Vehicles

FM 5–20C, Camouflage of Bivouacs, Command Posts, Supply Points, and Medical Installations.

d. Since ships being unloaded are likely to be targets for enemy aircraft, DUKW's lying alongside should, when ordered to do so, cast off immediately when enemy planes are sighted. Each DUKW will then proceed as far away from the ship as possible, steering a zig-zag or roundabout course. DUKW's that are in the water close to the beach should land and get under cover.

e. Warning of aircraft attacks must be received in time to be effective. Unit commanders should make definite and early arrangements with shore aircraft warning services to furnish such information to the unit. Warnings also should be received by arrangement with the ship's personnel. A predetermined signal on the ship's whistle or the use of parachute flares will indicate that the warning has been received. Audible signals are difficult to hear during operations. The assistant driver therefore must be constantly on the alert for air attacks and for all warning signals.

124. ANTIMECHANIZED DEFENSE. a. Antimechanized measures for an amphibian truck unit are similar to those employed by other units. (See FM 100–5.) The weapons available and the operations recommended for antiaircraft defense can also be used to advantage against mechanized forces. In dispersing or taking evasive action, care must be taken to avoid mine fields. Prompt and accurate information with respect to mines is essential.

b. Islands in rivers and offshore which are inaccessible to mechanized units should be utilized as bases of operations for the amphibian truck unit. Crowding should be avoided and routes of approach must be reconnoitered prior to the occupation of islands. These bases rarely will be out of range of the weapons of mechanized units. Shelter for personnel and vital equipment should be constructed without delay. Ground patrols and outposts must be established at once to give adequate warning of enemy attack. Full advantage should be taken of deep sand dune areas for protection against hostile mechanized forces.

125. DISPERSAL OF VEHICLES. Due consideration should be given to the dispersal of DUKW's consistent with the tactical situation. DUKW's should not be concentrated in the water waiting to come along side of the ship, nor should they be concentrated in the dump or dispersal areas.

KNOTS, BENDS, FENDERS AND NETS

---

126. KNOTS AND BENDS. Knots, bends, whipping, and splicing are used constantly in DUKW operations. Frequently the safety of the DUKW and its crew will depend on the ability to make a dependable knot quickly. DUKW crews should be familiar with the knots, bends, whipping, and splicing shown in this chapter. See TM 5–225 and TM 55–310 for additional information.

a. Square knot. For square knot see figure 71.

Figure 71. Square knot.

This knot is suitable for joining two ropes of approximately the same size. If a square knot is pulled hard it will not come undone, but will be difficult to untie.

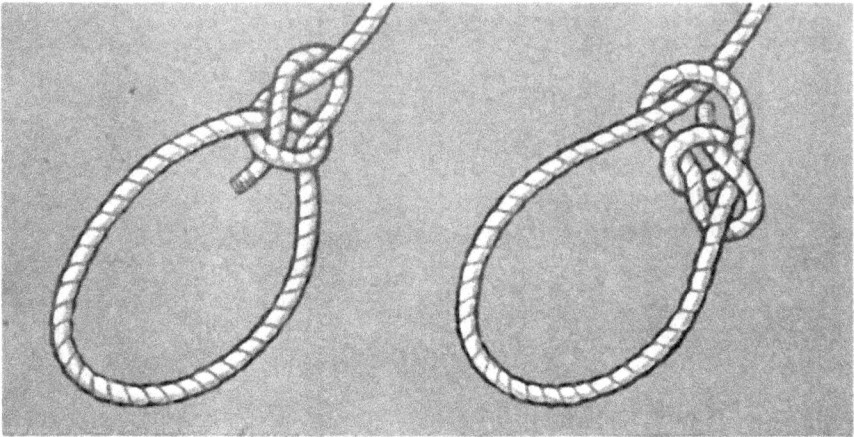

Figure 72. Bowline and running bowline.

b. Bowline. For a bowline see figure 72. It will hold very well and is the easiest knot to undo after a heavy strain. The rope, however, must be slack when it is untied.

c. Running bowline. For a running bowline see figure 72. It makes a useful slip knot, formed by using a bowline and tying a small loop around the main part of the same rope.

d. Rolling hitch. The rolling hitch is used on the tarpaulin. It is also handy for mooring lines when a knot must be tied in a rope that is already tight. The rope should be wrapped twice around whatever it is to be tied to (see *a* in fig. 73) and then tied back on itself as shown in *b* in figure 73.

Figure 73. Rolling hitch will frequently be found useful.

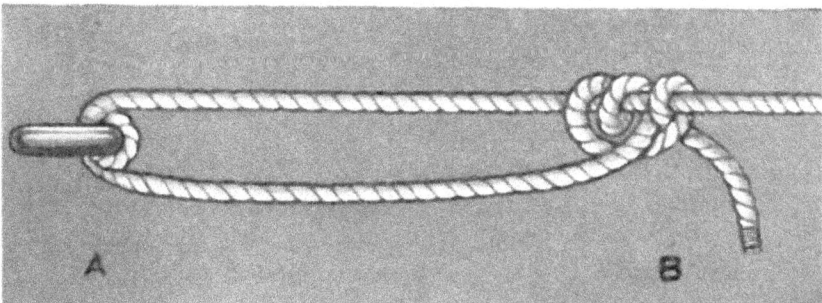

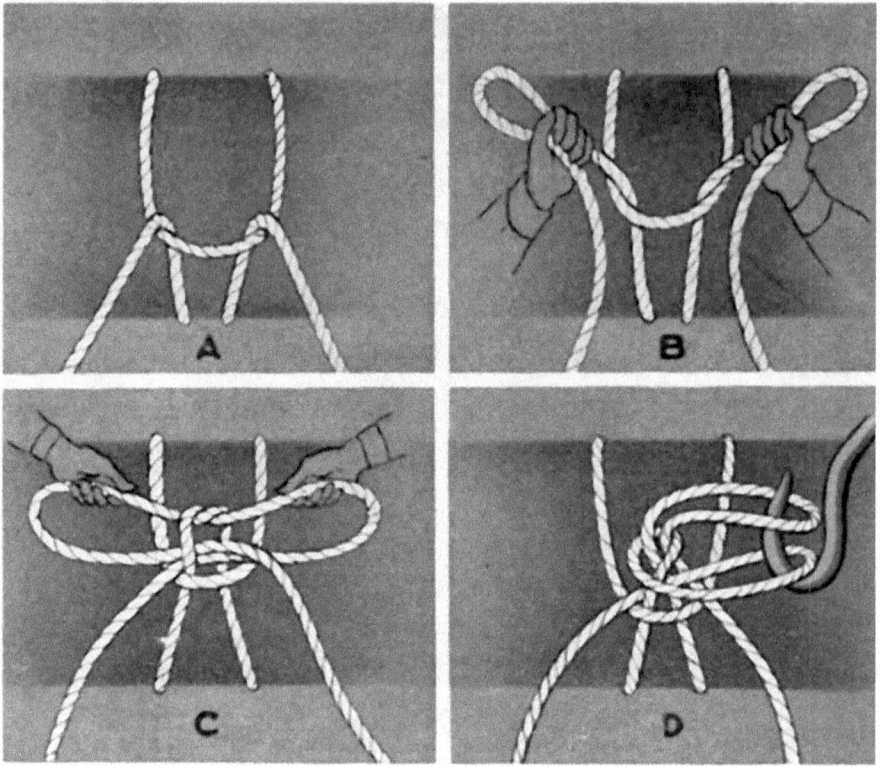

Figure 74. Shortening cargo sling with a sling knot.

e. Sling knot. The sling knot is used to shorten a continuous sling that is around a package which is to be lifted. (See fig. 74.) A short bight of the excess sling length is held in each hand (see *b* of fig. 74), and the double ropes forming those two short bights are tied in an overhand knot (see *c* of fig. 74) like the first half of a square knot. The hook for lifting the package is then hooked through the two ends of the short bights as shown in *d* of figure 74.

f. Whipping. Whipping a rope means to wrap a light twine around the end of the rope to prevent the rope fraying or unwinding. (See fig. 75.) Start about 1 inch from the end, the light twine or marline is layed in a loop and then the twine is wrapped around the loop and the

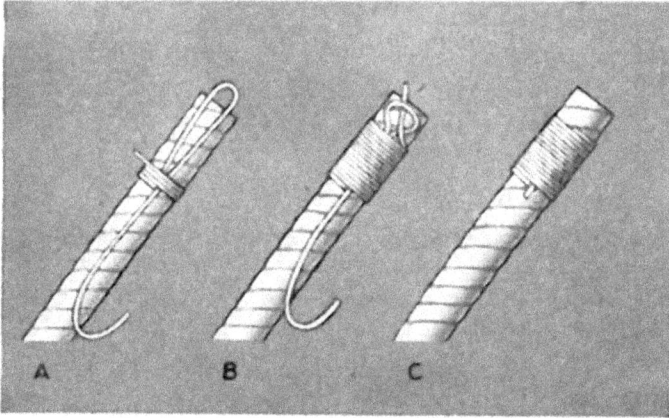

Figure 75. Whipping a rope.

rope. (See *a* of fig. 75.) When a point about ¼ inch from end of rope is reached, the end is led through the remaining part of the original loop. (See *b* of fig. 75.) This end is pulled *half way* through the whipping, and the original tail which was used to pull the other end down is cut off.

g. Short splice. The short splice is used to join two ropes together permanently. (See fig. 76.) Strands from each rope should be tucked three times into the other rope, and cut off the ends. *D* in figure 76 shows splice at completion of first tuck. The three strands on each end must each be tucked two more times.

Figure 76. Starting a short splice.

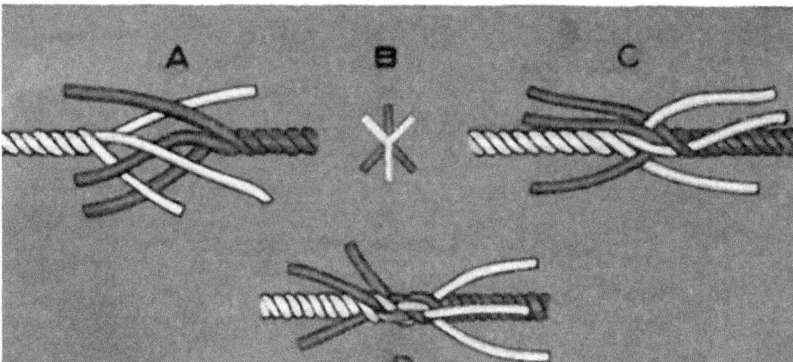

h. Eye splice. The eye splice is used to make a permanent loop in a rope. Each strand must be tucked three times. Figure 77 shows first tuck of first, second, and

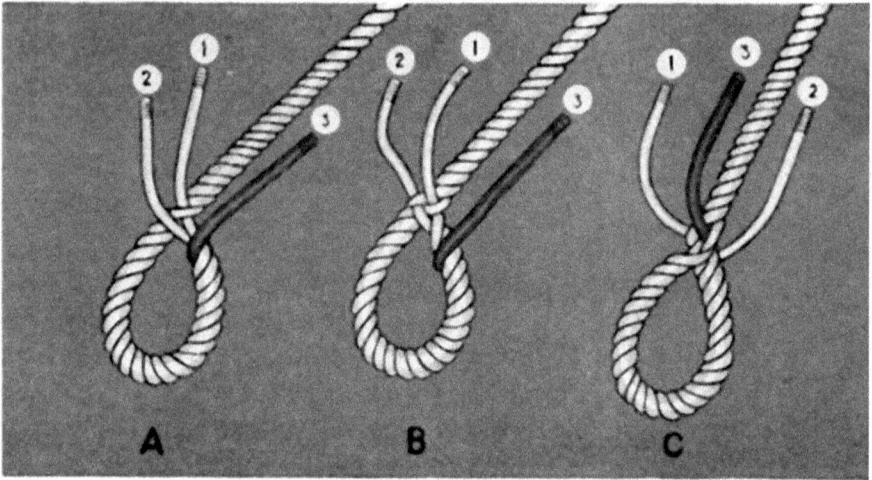

Figure 77. Starting and eye splice.

third strands, with two full tucks to make before splice will be completed.

i. Wire eye. Wire eye with **U**-clamps is used instead of eye-splicing a wire rope. Two or three clamps should be used, with base of clamps against main part of wire as

*Figure 78. Wire eye with **U** clips.*

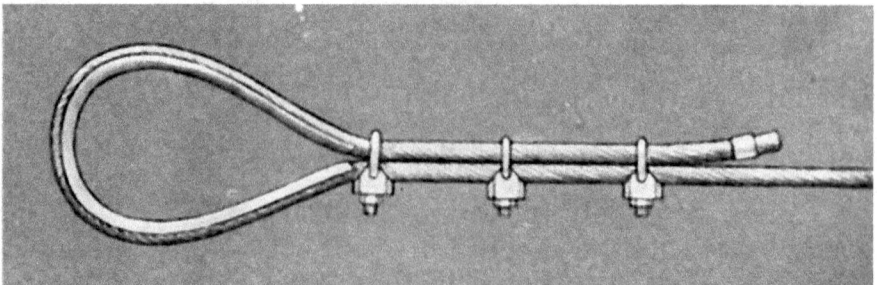

shown in figure 78. The clamps should be tightened evenly, and retightened after the initial strain has come on the wire.

127. HOW TO MAKE FENDERS. Fenders, sometimes called bumpers, are rope cushions used to protect the DUKW when it moors alongside a ship or wharf. In rough water operations, the fender will wear out rapidly. It is important, therefore, that DUKW companies become familiar with the details of making fenders. Fenders are made as follows:

a. The first step, shown in *A* of figure 79, is to make the core which should be 40 inches long over-all, including eyes. The center part of the core is made from two pieces of line, each of which is doubled, and has a 4-inch eye placed in each end. These two center parts should be lashed together securely at several points. The diameter of the center or core should be increased to about 5 inches by adding ropes wrapped with canvas or rags securely bound in place with twine or marline, as shown in *C*.

b. The next step, which is to hang the core on a hook at the highest convenient working height, is shown in *B*. Three lines are then led through the upper eye, with the center of the lines resting in the eye. These lines should be rope of $3/8$ to $1/2$ inch in circumference, each one measuring about 35 feet in length. (See figure 79.)

c. For purpose of description, the ropes are numbered from 1 to 6, starting with the front left rope and going in a counterclockwise direction. Ropes Nos. 1 and 6, 2 and 5, and 3 and 4, are opposite halves of the same ropes, respectively. No. 6 is brought around the left side of the core and under No. 1, leaving a small loop. Now No. 1 is led under 6, at the same time extending No. 1 to the right. Holding No. 6 in the left hand, No. 2 is carried over in front of No 1. Then No. 1 is led over No. 2 toward the left and into the left hand, completing C.

d. This same procedure is continued by bringing No. 3 over in front of No. 1. Then No. 1 is led over No. 2 toward the left and into the left hand. No. 4 next comes over in front of No. 3 and No. 3 over No. 4 toward the left and into the left hand. Next, No. 5 is brought over No. 4 and No. 4 over in front of No. 5 toward the left into the left hand. No. 5 is then led around in back of the core and through the small loop under 6 as shown in D of figure 79. (This loop was made when No. 6 was brought around the left side of core and under No. 1 as the first step in c above.)

e. E shows the completion of the first row. Each of the 6 ropes has been tucked over and all pulled up evenly and tightly.

f. F shows the fender after four full rows have been completed and the fifth row is having its final tuck preparatory to tightening. After each round all ends hanging down must be untangled and kept clear.

g. The above procedure is continued until the lower eye is reached as shown in G.

h. At that point, the fender is taken down and turned bottom end up. It is held between the knees, and two pairs of ropes are criss-crossed through the eye as shown in H.

i. The crown is completed by leading the remaining pair of ropes as shown in I. It must be noted that the last pair of ropes do not go through the eye.

j. J shows how to pull the crown down tightly, always pulling on two pair of ropes at a time.

k. The fender is hung in the original position and, using a marline spike or fid as shown in K, the strands are tucked back in a manner similar to splicing.

158

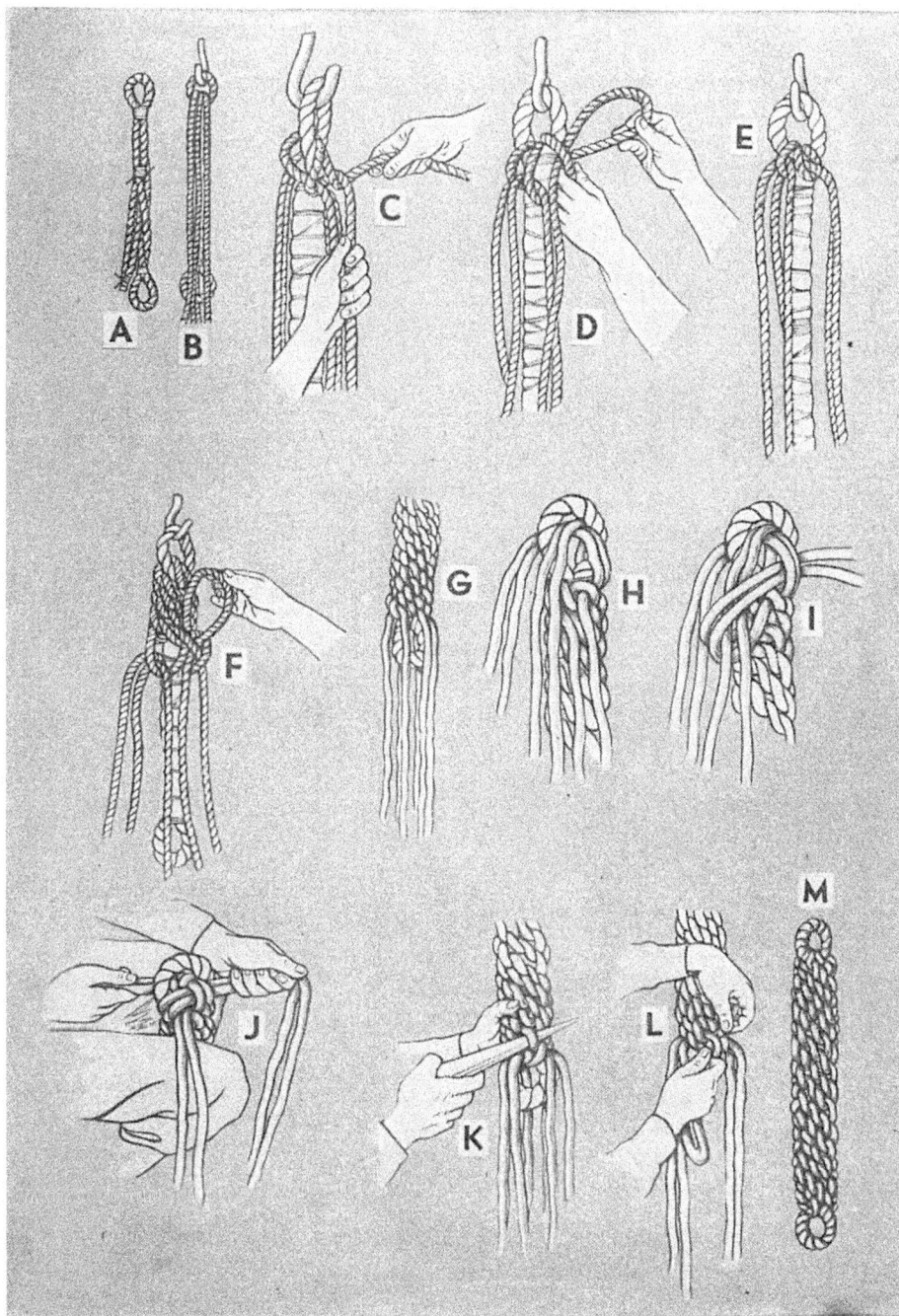

Figure 79. Making a fender.

CROSS TYPE CARGO NET

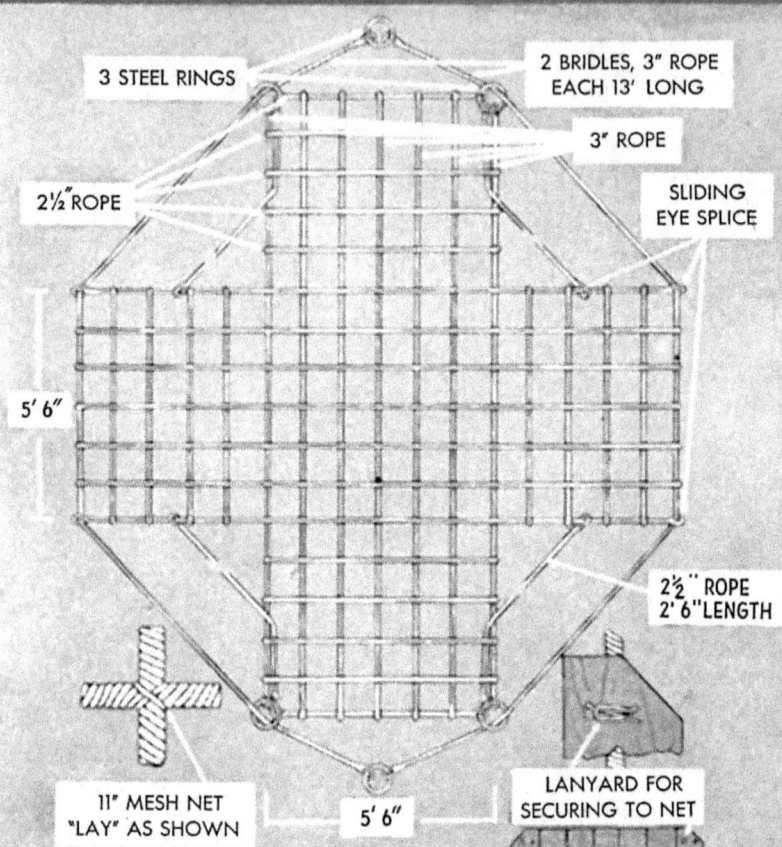

3 STEEL RINGS

2 BRIDLES, 3" ROPE EACH 13' LONG

3" ROPE

SLIDING EYE SPLICE

2½" ROPE

5' 6"

2½" ROPE 2' 6" LENGTH

11" MESH NET "LAY" AS SHOWN

5' 6"

LANYARD FOR SECURING TO NET

FINISHED LENGTHS
No Allowance Made for Splicing

14 lenghts of 3" Cir. Rope, each 13' 6" long
18 lengths of 2½" Cir. Rope, each 5' 6" long
4 lengths of 2½" Cir. Rope, each 2' 6" long
2 lengths of 3" Cir. Rope, each 13' long
6 Steel Rings; 4" I. D. and 5" O. D.

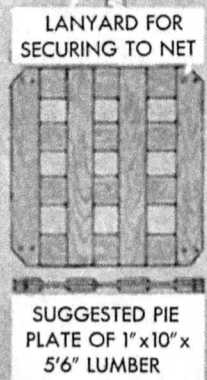

SUGGESTED PIE PLATE OF 1" x 10" x 5'6" LUMBER

Figure 80.

160

l. Each rope should be tucked under one strand, over the next and under the next, against the lay of the fender, as shown in L.

m. Figure M shows complete fender after loose ends have been pulled tight and then cut off.

128. NEW CROSS TYPE CARGO NET. a. General.
A new type of cargo net, shown in figure 80, has been developed combining the advantages of the safety of a regular net with the capacity and cargo protection of a pallet. This new type net is well suited to DUKW operations.

Figure 81. Cross type cargo net lifted with spider hooks by an A-frame.

b. On shipboard, where there is ample height to the cargo booms, the net should be picked up by hooking on the two eyes in the long bridle. For loading or unloading with short crane or **A**-frame, a four-legged hook-bridle should be used, as shown in figure 81. In transit, the long bridle is tied across the top to hold all the cargo in place.

SECTION XVIII

DESTRUCTION OF EQUIPMENT

---

129. DESTRUCTION OF MATÉRIEL IN EVENT OF IMMINENT CAPTURE. a. Tactical situations may arise when, owing to limitations of time or transportation, it will become impossible to evacuate all equipment. In such situations it is imperative that all matériel which cannot be evacuated be destroyed to prevent its capture and use by the enemy.

b. The destruction of matériel, subject to capture or abandonment in the combat zone, will be undertaken only when, in the judgment of the military commander concerned, such action is necessary.

c. If possible, all machine guns mounted in DUKW's prior to their destruction will be detached and salvaged.

d. All echelons will be trained in effective destruction of matériel issued to them.

130. METHODS OF DESTRUCTION. a. General. Destruction will be as complete as available time, equipment, and personnel will permit. If thorough destruction of all parts cannot be completed, the most important features of the matériel will be destroyed. Parts essential to the operation or use of the matériel and those which cannot be easily duplicated will be destroyed or removed. The same essential parts will be destroyed on all like units to prevent construction by the enemy of one complete unit from several damaged ones by "cannibalization."

b. Choice of methods. The destruction methods outlined below are arranged in order of effectiveness. Destruction should be by the first method suggested if possible. If this cannot be used, one of the other methods will be followed in the priority shown. Whichever method is used, the sequence outlined will be followed to assure uniformity of destruction.

c. Command decisions. Certain of the methods outlined require special tools and materials, such as TNT and incendiary grenades, which may not be items of normal issue. The issue of such special tools and material, the vehicles for which issued, and the conditions under which destruction will take place are command decisions in each case, according to the tactical situation.

131. DESTRUCTION OF VEHICLES. a. By sinking.

Where deep water is available close to shore, the simplest method of destroying the DUKW is to run it offshore and sink it, keeping a boat or DUKW to get the crews back to shore. In no case will a DUKW be destroyed by sinking in water less than 50 feet deep because of the ease of salvage. The sinking may be done in one of two ways:

(1) The four drain valves on DUKW's equipped with them (above serial number 353–2005) should be opened at sea. The DUKW's will sink in from 5 to 10 minutes after the engine has stopped running.

(2) DUKW's not having drain valves (below serial number 353–2005) should be run out into the water with one of the large hull plugs removed. This may be done only if the vehicle does not have to go too far to reach the necessary deep water. Otherwise, one of the methods outlined below must be used.

b. By TNT charge. Portable fire extinguishers are first removed and emptied and the fuel tank punctured. A

2-pound charge of TNT is placed on top of the clutch housing and, if time permits, 2 pounds placed as low as possible on the left side of the engine. Tetryl nonelectric caps, with at least 5 feet of safety fuse, are inserted in each charge. The fuses should be ignited from outside the vehicle while personnel should take cover quickly, since gasoline fumes may be exploded prematurely by burning fuses. Elapsed time will be 1 to 2 minutes if charges are prepared beforehand and carried in the vehicle.

c. By gunfire. Portable fire extinguishers are first removed and emptied, the fuel tank punctured, and, if time is available, all doors and hatches opened. The vehicle is then fired on, using tank, antitank, or other artillery, or antitank rockets or grenades. The aim is directed at the engine. If a good fire is started, the vehicle may be considered destroyed. Elapsed time: About 5 minutes per vehicle.

d. By smashing units. Portable fire extinguishers are removed and emptied and the fuel tanks punctured. All vital parts (such as distributor, carburetor, radiator, engine block, air and oil cleaners, generator, control levers, crankcase, and transmission) should be smashed with a heavy axpick, or sledge. Spare gasoline, oil, or other inflammable liquid is poured over entire unit and ignited.

132. DESTRUCTION OF TIRES. a. General. Rubber is such a critical item that, whenever matériel is subject to capture or abandonment, an attempt to destroy pneumatic tires, including the spare, must always be made even if time will not permit destruction of any other part of the DUKW. With adequate planning and training, however, the destruction of tires may be accomplished in conjunction with destruction of the vehicle without increasing the time necessary.

b. By incendiary grenades. An M14 incendiary grenade is ignited under each tire. When this method is combined with the destruction of DUKW's by TNT, the incendiary fires must be well started before the TNT is detonated.

c. By damaging and burning. Tires may be damaged with an axpick, or by heavy machine-gun fire, deflating them first, if possible. Spare gasoline may be poured on tires, dousing each one, and ignited.

133. DESTRUCTION OF SHOP EQUIPMENT. a. Hand tools. (1) Though tools in general are not difficult to destroy, they are so numerous that effective destruction may require considerable time. Articles such as cold chisels are difficult to destroy to an extent which does not allow reclamation. Destructive efforts should be concentrated on delicate instruments.

(2) Where DUKW's are to be burned, the hand tools and machine tools should be left in them to be burned also.

b. Electric motors and generators. To destroy electric motors and generators, a cold chisel may be used to cut a number of turns of the windings. The armature should be destroyed first since armature coils are more difficult to replace than field coils. If the outside protective covering of insulation is broken, contact with acid will short circuit the connection and corrode the wires. If a cutting torch is available, the projecting end of the shaft should be cut off flush with the bearing.

c. Welding equipment. Since welding apparatus may be used to destroy other equipment, its destruction should be left to the last.

(1) Acetylene torches may be destroyed by striking threaded parts and union seats with a hammer, burning the hose, and smashing regulators. All gas should be let out

of the cylinders, and the valves and threaded connections broken.

(2) On electric welding generators, the coils should be cut and the end housings broken with a heavy hammer if possible. The machine should be burned with the DUKW.

134. DESTRUCTION OF DOCUMENTS. All Field Manuals, Technical Manuals, Standard Nomenclature Lists, and other documents should be burned.

APPENDIX I

COMBAT SHIPMENT OF DUKW'S

I. GENERAL. The following instructions should be followed where DUKW's are attended by crews during shipment and the DUKW's are to be put into service immediately on the arrival of the transporting ship at its destination. This is different from normal export shipping where the vehicle must be thoroughly cleaned and serviced before operating.

2. PAYLOAD. All instructions contained in paragraphs 84 to 90 regarding payloads apply with equal force in combat shipments of DUKW's. When DUKW's are launched from ships, the total payload should never exceed 7,500 pounds.

3. CARE OF DUKW'S WHEN CARRIED IN DAVITS.
a. The davit lifting eyes on the DUKW are spaced 22 feet 5 inches apart. If the measurement between the davit heads on the ship differs from this spacing, either a spreader bar must be used or a lashing put on the falls, in order to keep the strain on the davit eyes straight up. This will, however, increase the strain on the falls.

   b. If the DUKW davit lifting eyes are not large enough to receive davit hooks, shackles can be fitted in the davit eyes. These shackles should be made to stand upright, unsupported, by wrapping marline or rope yarns around them. This will greatly facilitate hooking and unhooking.

In many instances, it will be helpful to slide the anchor 5 or 6 inches toward the stern to clear the rear davit lifting eye. When this is done, the anchor lashing straps should not be removed but merely loosened so that the anchor cannot slide off the rear deck. In some cases, where the davit hook will not fit in the davit eye, removal of the caution plate on the deck will provide sufficient additional room so that the hook will fit.

c. A DUKW with standard equipment, but *without payload,* will weigh not more than 7,750 pounds on the after davit hook and 7,250 pounds on the forward davit hook. The davit lifting eyes are designed to lift safely a DUKW loaded with the rated payload of 5,000 pounds. The payload in the cargo space will fall about two-thirds on the after davit hook and one-third on the forward davit hook.

d. When the DUKW is overloaded for an emergency operation, additional auxiliary slings must be attached to the DUKW in such a manner that exceptional loads are not put on any of the davit or regular lifting eyes.

e. A two-legged transverse bridle of not less than 5/8-inch diameter plow steel wire rope *can be* attached to the side lifting eyes for the purpose of hoisting the DUKW if davit lifting eyes are not installed or have been carried away. When two-legged bridles are used, each leg should be at least 8 feet long.

f. When DUKW's are carried outboard, *chocking blocks* should be fitted against the full height of the DUKW's side. The block face should be at least 6 inches wide and be padded to protect the paint. The blocking should be so arranged that it will not press against the light metal shields which cover the wheels. Griping should be run from the center forward shackle and after towing pintle. Gripes leading completely around the hull of the DUKW and gripes leading to davit eyes should be avoided. The cast iron fairlead in the bow is not strong enough to use for lashing.

g. To make sure that the DUKW will be in operating condition upon arrival at the destination, the operator should—

(1) Paint over the eight hull reflectors.

(2) On the DUKW's so fitted, ream the hinges for the plywood rear closure, to help removal and reinstallment.

(3) Perform daily and weekly maintenance, omitting only those items that conditions make obviously unnecessary or impossible.

(4) On vehicles so equipped, leave the four drain valves open. Since, when the DUKW is swung outboard, no bottom or housing plugs can be removed or replaced, water must be sucked out of shaft housings from inside. During freezing weather, check carefully to make sure that these housings do not contain water.

(5) Run engine 15 minutes each day, operating the propeller and all wheels in all gear combinations for a short time. In freezing weather, check to see that bilge pumps are free by turning the propeller shaft by hand. Operate the air pump, winch, and brakes, and put the steering gear hard over each way several times. Coat the front axle universal joint housings with oil.

(6) Cover the vehicle tightly and clamp all deck hatches shut.

h. Before the DUKW's are put overboard at the landing place, drivers should check those items in the check list, paragraph 7 below, which apply.

4. CARE OF DUKW'S WHEN CARRIED ON DECK. a.
The DUKW's are hoisted onto the deck with the aid of a plow steel wire bridle at least 5/8-inch diameter, with four

equal legs at least 16 feet long. The bridle is attached to the four side lifting eyes of the DUKW, with suitable hooks or shackles which must be sufficiently strong to lift the DUKW, fully loaded.

b. When the DUKW's are stowed on deck, their tires should be inflated to 40 pounds, and chocked. The space between the axles and the frame should not be blocked. The lashings should be made fast to the three bow shackles and to either the stern pintle or the two eyes on the rear corners. The four lifting eyes and two mooring eyes may also be used if the lashings are kept close to the sides to avoid side pull. The bow fairlead or the two davit eyes should not be used. (See fig. 82.) The driver should shift the gears to low-range reverse and set the hand brake.

c. To make sure that the DUKW will be in operating condition on arrival at the destination, the operator should—

(1) Paint over the eight hull reflectors.

(2) Perform daily and weekly maintenance, omitting only such items as conditions make obviously unnecessary.

(3) Run the engine 15 minutes daily. In freezing weather, check that bilge pumps are free by turning propeller by hand. Operate propeller, air pump, and winch for short periods. Turn the steering gear from hard over to hard over if the wheel blocking permits. Coat the front axle universal joints housings with oil.

(4) Clean the bilges without removing the three large hull bottom plugs. The four hull drain valves should be left open. Drive shaft housings should be drained occasionally as necessary, and plugs left in place, except in freezing weather when plugs should be removed and tied to the steering wheel or placed in the rack in the windshield base.

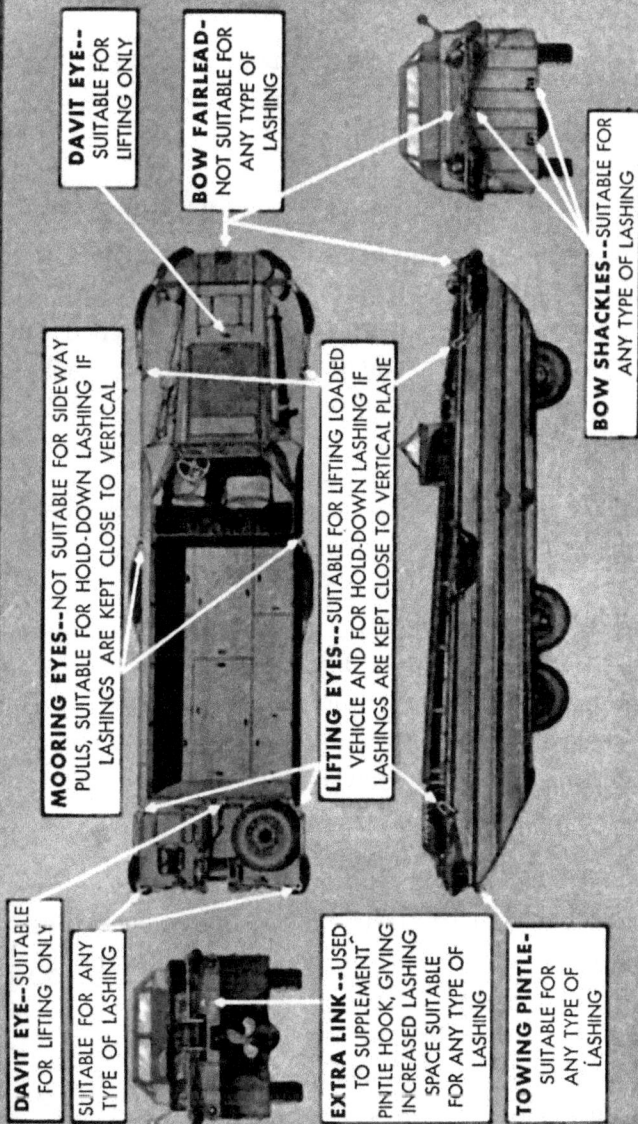

LASHING INSTRUCTIONS

DAVIT EYE-- SUITABLE FOR LIFTING ONLY

BOW FAIRLEAD- NOT SUITABLE FOR ANY TYPE OF LASHING

BOW SHACKLES--SUITABLE FOR ANY TYPE OF LASHING

MOORING EYES--NOT SUITABLE FOR SIDEWAY PULLS, SUITABLE FOR HOLD-DOWN LASHING IF LASHINGS ARE KEPT CLOSE TO VERTICAL

LIFTING EYES--SUITABLE FOR LIFTING LOADED VEHICLE AND FOR HOLD-DOWN LASHING IF LASHINGS ARE KEPT CLOSE TO VERTICAL PLANE

DAVIT EYE--SUITABLE FOR LIFTING ONLY

SUITABLE FOR ANY TYPE OF LASHING

EXTRA LINK--USED TO SUPPLEMENT PINTLE HOOK, GIVING INCREASED LASHING SPACE SUITABLE FOR ANY TYPE OF LASHING

TOWING PINTLE- SUITABLE FOR ANY TYPE OF LASHING

Figure 82.

173

(5) Cover the vehicle tightly and clamp all deck hatches shut.

d. Before the DUKW's are put overboard at the landing place, drivers should check these items in the check list, paragraph 7 below, which apply.

5. CARE OF DUKW'S WHEN CARRIED IN HOLD.

a. The DUKW's are hoisted into the hatch with the aid of a plow steel wire bridle at least $\frac{5}{8}$-inch diameter with 4 equal legs at least 16 feet long. The bridle is attached to the four side lifting eyes of DUKW, with suitable hooks or shackles which must be sufficiently strong to lift the DUKW, fully loaded.

b. When the DUKW's are stowed in the hold, their tires should be inflated to 40 pounds, and chocked. The space between the axles and the frame should not be blocked. The lashings should be made fast to the three bow shackles and to the stern pintle or the two eyes on the rear corners. The four lifting eyes and two mooring eyes may also be used if the lashings are kept close to the sides to avoid side pull. The bow fairlead or the two davit eyes should not be used. The driver should shift the gears into low-range reverse and set the hand brake.

c. To make sure that the DUKW will be in operating condition upon arrival at destination, the operator should, before the DUKW is stowed in the hold:
(1) Paint over the eight hull reflectors.
(2) Perform daily "after operation" maintenance.
(3) Disconnect battery wires.
(4) Open four hull drain valves.
(5) Cover the vehicle where possible and clamp all deck hatches shut.

d. Before the DUKW's are put overboard at the landing

place, drivers should check those items in the check list in paragraph 7 below, which apply.

6. CARE OF DUKW's ON LANDING CRAFT WITH RAMPS.

a. Where time permits the DUKW's should be *backed* aboard, to expedite unloading at destination. Ramps must be clear of sharp projections that would damage hull or tires.

b. When the DUKW's are on deck or in the hold, their tires should be inflated to 40 pounds, and chocked. The space between the axles and the frame should not be blocked. The lashings should be made fast to the three bow shackles and to the stern pintle or the two eyes on the rear corners. The four lifting eyes and two mooring eyes may also be used if the lashings are kept close to the sides to avoid side pull. The bow fairlead or the two davit eyes should not be used. The driver should shift the gears to low-range reverse and set the hand brake.

c. To make sure that the DUKW will be in operating condition on arrival at the destination, the operator should—

(1) Paint over the eight hull reflectors.

(2) Perform daily and weekly maintenance, omitting only such items as conditions make obviously unnecessary.

(3) Run the engine 15 minutes daily. In freezing weather, check that bilge pumps are free by turning propeller by hand. Operate propeller, air pump, and winch for short periods. Turn the steering gear from hard over to hard over if the wheel blocking permits. Coat the front axle universal joint housings with oil.

(4) Clean the bilges without removing the three large hull bottom plugs. The four hull drain valves should be left open. Drive shaft housings should be drained occasionally as necessary and plugs left in place, except in freezing weather—when plugs should be removed and tied to the

steering wheel, or placed in the rack in the windshield base.

(5) Cover the vehicle, and clamp all deck hatches shut.

d. Before the DUKW's are launched from the landing craft, drivers should check those items in the check list, paragraph 7 below, which apply.

7. BEFORE LAUNCHING CHECK LIST. a. "Before operation" maintenance accomplished.

b. Tire pressure.

c. Supply of fuel, oil, and water.

d. Battery connected and charged.

e. Weight of payload.

f. Cargo properly placed and secured.

g. Bottom plugs and valves.

h. Bow surf plate.

i. Windshield surf plate.

j. Deck hatches clamped shut.

k. Cargo tarpaulin rigged if required.

l. Forward bilge pump opened.

m. Engine cooling system clear and adjusted.

n. Unhooking instructions understood by crew (aft fall or legs of bridle first, necessity for teamwork and speed).

o. Messenger line attached to davit fall or bridle leg to clear after unhooking.

p. Line secured to inboard bow lifting eye to be led forward on ship to hold DUKW while unhooking.

q. Crew aboard.

r. Propeller transfer case engaged.

s. Engine warmed up.

t. Hand brake released.

APPENDIX II

PERTINENT REFERENCES

MANUALS

FM 5–20, Camouflage, Basic Principles.

FM 5–20B, Camouflage of Vehicles.

FM 5–20C, Camouflage of Bivouacs, Command Posts, Supply Points, and Medical Installations.

FM 8–40, Field Sanitation.

FM 9–39, Bomb Handling.

FM 21–10, Military Sanitation and First Aid.

FM 21–11, First Aid for Soldiers.

FM 21–22, Watermanship.

FM 21–25, Elementary Map and Aerial Photograph Reading.

FM 21–30, Conventional Signs, Military Symbols, and Abbreviations.

FM 21–40, Defense Against Chemical Attack.

FM 23–65, Browning Machine Gun Caliber .50, HB M2 (mounted in combat vehicles).

FM 25–10, Motor Transport.

FM 31–5, Landing Operations on Hostile Shores.

FM 55–130, Small Boats and Harbor Craft.

TM 5–525, Rigging and Engineer Hand Tools.

TM 5–1174, Crane, Truck-mounted, (Quickway Model E).

TM 9–802, Truck, Amphibian, 2½-ton, 6 x 6 (GMC, DUKW–353).

TM 9–1802A, Power Plant for 2½-ton Amphibian Truck, 6 x 6 (GMC, DUKW–353).

TM 9–1802B, Power Train for 2½-ton Amphibian Truck, 6 x 6 (GMC, DUKW–353).

TM 9–1802C, Hull and Water Drive for 2½-ton truck, 6 x 6 Amphibian, (GMC, DUKW–353).

TM 10–460, Driver's Manual.

TM 55–310, Stevedoring.

FILMS AND FILM STRIPS

FB 60, DUKW's, The Seagoing Truck.

TF 9–1328, Truck, Amphibian, 1st Echelon Maintenance, Part I.

TF 9–1329, Truck, Amphibian, 1st Echelon Maintenance, Part II.

TF 55–1117, Military Stevedoring, Part IV, Drafts and Slings.

TF 55–1118, Military Stevedoring, Part V, Straps and Bridles.

TF 55–1119, Military Stevedoring, Part VI, Vehicle Loading and Stowing.

FS 9–212, Truck, Amphibian, Preventive Maintenance, First Echelon, Part I.

FS 9–213, Truck, Amphibian, Preventive Maintenance, First Echelon, Part II.

FS 9–214, Truck, Amphibian, Preventive Maintenance, First Echelon, Part III.

FS 9–215, Truck, Amphibian, Preventive Maintenance, First Echelon, Part IV.

FS 9–216, Truck, Amphibian, Preventive Maintenance, Removal of Engine, Part I.

FS 9–217, Truck, Amphibian, Preventive Maintenance, Removal of Engine, Part II.

FS 9–218, Truck, Amphibian, Preventive Maintenance, Second Echelon, Part I.

FS 9–219, Truck, Amphibian, Preventive Maintenance, Second Echelon, Part II.

FS 9–220, Truck, Amphibian, Preventive Maintenance, Removal of Amphibian Units, Part I.

FS 9–221, Truck, Amphibian, Preventive Maintenance Removal of Amphibian Units, Part II.

APPENDIX III

GLOSSARY OF SEA TERMS AND WORDS

Abeam. An object at right angles to the center line of the craft is said to be "abeam."

Accommodation ladder. Steps slung at the gangway leading down the ship's side to a point near the water, for boarding the ship from small boats.

Aft. At, toward, or near the stern, or rear, of a craft.

Amphibian truck. A wheeled, self-propelled vehicle designed to operate on either land or water.

Amidships. At, toward, or near the center of a craft.

Anchor. A device equipped with two hooks, or flukes, used to hold a craft in one place in the water. Not to be confused with *sea anchor.* When used on land in winching operations, it is called a *sand anchor.*

Astern. An object behind, or off the rear of a craft is said to be "astern"; also the direction of moving in reverse.

Athwartship. Across a vessel; from side to side.

Beam. The width of a craft at its widest part.

Bearing. The direction of an object from a ship.

Berth. A place where vessels dock, moor, or anchor; a bed or place to sleep.

Between decks. The space between any two decks, not necessarily adjacent. Frequently expressed as "tween decks."

Bight. A loop or bend in a rope; strictly, any part between two ends of a rope.

180

Bilge. The interior bottom of a craft.

Bitter end. The inboard end of a vessel's anchor chain which is made fast in the chain locker.

Bitts. Cast steel heads serving as posts to which cables are secured on a ship.

Blinker lights. Two electric lanterns secured at the ends of the signal yard and operated by controllers and a telegraph key for use in night signaling by code.

Block. A shell, housing one or more sheaves over which a cable is passed; used to increase pulling power.

Boom. A heavy pole or spar used in cargo handling.

Bow. The forward part of a craft.

Bow lines. A rope leading from the vessel's bow to another vessel or to a wharf for the purpose of hauling her ahead or for securing her.

Buoy. A floating body moored to the bottom of a body of water to show position of rocks, shoals, or to indicate a channel.

Breast line. A line running directly across a craft at front or rear, as used when mooring alongside.

Bridge. A high transverse platform, often forming the top of a bridge house, extending from side-to-side of the ship, and from which a good view of the weather deck may be had.

Bridle. A piece of rope each end of which is made fast to an eye forming a loop. Used in a bridle tow.

Broached boat. A boat tossed up on the beach, frequently the result of wind and surf pounding.

Bulkhead. A partition or wall separating compartments.

Cannibalization. Act or practice of taking good parts from matériel that is beyond repair and using them to repair other matériel.

Cargo hatch. A large opening in the deck to permit loading of cargo.

Cargo net. A net, made in various sizes, of manila or wire rope, used in connection with the vessel's hoisting appliances to load cargo, etc., aboard the vessel.

Chock. To brace cargo with dunnage. To block wheels of vehicles to prevent rolling.

Cleats. Pieces of wood or metal, of various shapes according to their uses, usually having two projecting arms or horns upon which to fasten ropes.

Coaming. A raised section or wall around hatches and other openings in the deck, to prevent the entry of water.

Compass. An instrument for showing direction. A magnetic compass has a magnetic needle that points to the magnetic north pole.

Cradle. A support of wood or metal shaped to fit the object which is stowed upon it.

Davit. A device used to lower and raise ship's boats or other objects.

Deadlined. A vehicle that is awaiting repairs.

Deadman. A buried log serving as a winching point for the winch cable in winching operations.

Deadweight tonnage. Actual carrying capacity of a vessel, in tons of 2,240 pounds, including stores, fuel, water, and cargo. (See TM 55–310.)

Dock. The water space alongside a wharf or pier for the berthing of vessels.

Draft. The vertical distance below the water surface of the deepest part of a craft.

Dunnage. Pieces of wood, mats, boughs, or loose materials of any kind, laid on the bottom of the cargo space or placed between cargo to prevent injury from motion and chafing, to give ventilation and to protect cargo from moisture.

Fairlead. A ring or opening, serving as a guide for rope or cable. On the DUKW, the fairlead is a fitting at the bow.

Fall. By common usage, the entire length of rope used in a tackle; sometimes limited in application to that end to which the power is applied. The end secured to the block is the standing part, the opposite end, the hauling or running part.

Fathom. A nautical unit of length used in measuring cordage, chains, depths, etc. The length varies in different countries, being 6 feet in the United States and in Great Britain.

Fenders. Cushions to protect the vessel coming alongside another vessel or wharf. They may be of wood, braided rope, old tires, or canvas filled with cork.

Fissure. A narrow opening made by the parting of coral or a rock.

Fore. Parts of a ship at or adjacent to the bow; also applied to parts of a ship lying between the midship section and stem, as fore body, fore hold, and foremast.

Forward. At, toward, or near the bow of a craft.

Freeboard. The verticle distance from the water surface to the top of the hull of a craft.

Gangplank. A movable platform used in transferring passengers or cargo from a vessel to a dock or wharf.

Gripe. An arrangement for holding a small boat securely in its stowage chocks.

Guest warp. A long rope cable running, at the water line, along the ship's side from bow to stern; used in mooring DUKW's to a vessel.

Guy. A rope attached to anything to steady it.

Hard over. The process of turning the steering wheel to the right or left as far as it will go.

Hatch.   An opening in the deck of a vessel for passage of cargo to the hold.

Hoist.   To raise or elevate; any device employed for lifting weights.

Hold.   That part of the interior of the hull in which cargo is stowed.

Hull.   The body of a watercraft.

Inboard.   Toward the center, within the vessel's shell and below the weather decks.

Keel.   The principal fore-and-aft member of a ship's frame. A center-line strength member running fore and aft along the bottom of a vessel and often referred to as the backbone. The keel runs along the bottom, connects the bow and stern, and serves as an anchor for the frames which are attached to it.

King post.   A strong vertical post used to support a derrick boom.

Knot.   A tie in a line. A unit of speed, equaling 1 nautical mile (6,080.20 feet) an hour.

Lanyard.   A piece of rope or line having one end free and the other attached.

Lee.   Away from the wind or weather side of a vessel; the sheltered side.

Lighter.   A large open barge used in loading and unloading vessels or in carrying freight around a harbor.

Line.   A general term for a rope.

List.   The leaning over to one side of a ship.

Locker.   A storage compartment.

Marlinespike.   A pointed iron or steel tool used to separate the strands in splicing rope, and as a lever in putting on seizings.

Marline.   A double-threaded, left-handed tarred cord, about 1/8 inch in diameter, made of a good grade of American hemp.

184

Mast. A long pole of steel or wood originally used for carrying sails and now used more as supports for rigging cargo and boat-handling gear, and wireless equipment.

Messenger line. A light line used in hauling in a heavier line at shipside.

Moor. To secure a vessel or craft to a shift or wharf.

Mooring lines. The chains or ropes used to tie up a vessel.

Offshore. Away from the shore.

On deck. In the open air on board ship.

Outboard. Away from the center toward the outside; outside the hull.

Parceling. Strips of canvas which are wound around ropes, following the lay and overlapping in order to protect rope from chafe and wear.

Pendant. A short rope or chain.

Pitching. The alternate rising and falling motion of a vessel's bow in a nearly vertical plane as she meets the crests and troughs of the waves.

Plimsoll mark. A mark painted on the side of a vessel designating the depth to which, under the maritime laws, the vessel may be loaded in different bodies of water during various seasons of the year.

Ponton. A scow-shaped vessel used in connection with engineering and military operations such as transporting men and equipment, bridge construction equipment, supports for temporary bridges, salvage work, etc.; also applied to cylindrical air and watertight tanks or floats used in salvage operations.

Poop. The structure or raised deck at the after end of a vessel.

Port. The left-hand side of a watercraft when looking toward the bow from the stern.

Porthole. A round window in the side or deck house of a vessel fitted with a hinged frame in which a thick glass is secured.

Propeller tunnel. Concave opening at the rear of the DUKW in which the water propeller and rudder are mounted.

Range markers. Markers set up on the beach generally in pairs to indicate the direction in which watercraft should be headed.

Reeve. To pass the end of a rope or chain through an opening.

Ride. To float in a buoyant manner.

Rigging. A term used collectively for all the ropes and chains employed to support the masts, yards, and booms of a vessel, and to operate the movable parts.

Roll. Motion of the ship from side to side.

Rudder. A device used in steering or maneuvering a vessel.

Rudderstock. A vertical shaft having a rudder attached to its lower end and having a yoke, quadrant, tiller, or lever fitted to its upper portion, by which it may be turned.

Running rigging. Ropes which are hauled upon in order to handle and adjust sails, yards, cargo, etc., as distinguished from standing rigging which is fixed in place.

Sea anchor. A strong, conical canvas bag, paid out at the end of a long cable; used to form a drag and reduce movement of a craft in water or to hold her bow into the sea during a storm. It is not a true anchor in that it does not sink to the bottom. In cases of emergency, a sea anchor can be improvised from various bulky objects on board.

Shackle bolt. A pin or bolt that passes through both eyes of a shackle and completes the link.

186

Sheave. The wheel inside a block around which a rope or cable runs.

Shipshape. Neat in appearance, and in good order.

Skeg. A projection of the ship's keel on which rests the lower end of the post to which the rudder is attached.

Slack. Not fully extended; to "slack away" means to pay out a rope or cable by carefully releasing the tension while still retaining control; to "slack off" means to ease up or lessen the degree of tautness.

Sling. A length of rope or chain employed in handling weights with a crane or davit.

Snatch block. A block in which the housing opens on one side to admit the rope or cable without threading it over the sheave.

Spar. A pole serving as a mast or a boom.

Splice. A method of uniting the ends of two ropes by first unlaying the strands, then interweaving them to form a continuous rope.

Spring line. A line running diagonally from front or rear of a craft when mooring alongside a vessel or wharf. It is also used when towing another craft alongside.

Starboard. The right side of a watercraft when looking toward the bow from the stern.

Steering gear. The wheels, leads, and fittings by which the rudder is turned.

Stern. The rear part of a craft.

Strut. A support or brace. On the DUKW the "**V**-strut" is a **V**-shaped support which holds the water propeller in alignment.

Surf. The waves of the sea as they break upon the shore.

Tackle. A system of blocks through which a rope or cable is led to increase pulling power.

Tarpaulin. A heavy canvas used to cover objects that require protection from the weather.

Taut. The condition of a rope, wire, or chain when under sufficient tension to cause it to assume a straight line.

Tiller. A horizontal bar at the head of the rudder post; used to turn the rudder in steering.

Ton. The weight of 2,000 pounds. (For the different kinds of tonnage used in shipping cargo, see TM 55–310.)

Wake. The track left by a vessel in the water.

Wharf. A fixed structure to which a vessel moors to load or discharge cargo and passengers.

Whipping. Turns of twine wound around the end of a rope to prevent its unlaying.

Winch. A hoisting or pulling machine used principally in the handling, hoisting, and lowering of cargo.

Winching point. Any fixed object to which the winch cable or rigging is attached during winching operations.

Windward. The side from which the wind is blowing.

Note. For a more complete glossary of nautical terms see TM 55–310 and FM 55–130.

INDEX

190

☆   U. S. GOVERNMENT PRINTING OFFICE: 1944—604263

* 9 7 8 1 9 3 7 6 8 4 5 7 0 *